The Tao of Microelectronics

The Tao of Microelectronics

Yumin Zhang

Southeast Missouri State University, Cape Girardeau, MO, USA

Morgan & Claypool Publishers

Copyright © 2014 Morgan & Claypool Publishers

All rights reserved. No part of this publication may be reproduced, stored in a retrieval system or transmitted in any form or by any means, electronic, mechanical, photocopying, recording or otherwise, without the prior permission of the publisher, or as expressly permitted by law or under terms agreed with the appropriate rights organization. Multiple copying is permitted in accordance with the terms of licences issued by the Copyright Licensing Agency, the Copyright Clearance Centre and other reproduction rights organisations.

Rights & Permissions
To obtain permission to re-use copyrighted material from Morgan & Claypool Publishers, please contact info@morganclaypool.com.

ISBN   978-1-6270-5453-9 (ebook)
ISBN   978-1-6270-5452-2 (print)
ISBN   978-1-6270-5673-1 (mobi)

DOI   10.1088/978-1-6270-5453-9

Version: 20141201

IOP Concise Physics
ISSN 2053-2571 (online)
ISSN 2054-7307 (print)

A Morgan & Claypool publication as part of IOP Concise Physics
Published by Morgan & Claypool Publishers, 40 Oak Drive, San Rafael, CA, 94903, USA

IOP Publishing, Temple Circus, Temple Way, Bristol BS1 6HG, UK

Contents

Preface

Microelectronics is a challenging course to many undergraduate students and is often described as *very messy*. First, it covers many kinds of electronic device; for example, there is a whole family of diodes. Second, the characteristics of these devices are also rather complicated, for example the *I–V* curves of transistors. Third, various circuit configurations can be constructed, such as CB, CC and CE BJT amplifiers. Fourth, there are strong couplings between different parts of a circuit, such as circuits with feedback. Fifth, there are many conflicting constraints in design, such as bandwidth, power consumption, linearity, stability etc. In short, this topic is often considered a 'black art'.

Before taking this course, all the students have learned *circuit analysis,* where basically all the problems can be solved by applying Kirchhoff's laws. In addition, most engineering students have also learned *engineering mechanics: statics* and *dynamics,* where Newton's laws and related principles can be applied in solving all the problems. However, *microelectronics* is not as *clean* as these courses. There are hundreds of equations for different circuits, and it is impossible to remember which equation should be applied to which circuit. One of the common pitfalls in learning this course is over-focusing at the equation level and ignoring the ideas (Tao) behind it. Unfortunately, these ideas are not summarized and emphasized in most microelectronics textbooks, though they cover various electronic circuits comprehensively. Therefore, most undergraduate students feel at a loss when they start to learn this topic. This book tries to illustrate the major ideas and the basic analysis techniques, so that students can derive the right equations easily when facing an electronic circuit.

In chapter 1, the basic electronic devices are introduced in a coarse grained way; i.e., their basic features are highlighted and the details are ignored. Although some descriptions are not rigorous, the gist of these devices can be grasped very easily. In addition, the general ideas of rectifier and amplifier circuits are discussed. In chapter 2, the *pn*-junction diode is covered together with a brief introduction to semiconductor physics. Actually, the *pn*-junction is a structure that exists in most semiconductor devices, such as the BJT and MOSFET. In chapter 3, BJT characteristics and amplifier circuits are covered. Although the BJT is no longer widely used in integrated circuits, it is still an important topic. In chapter 4, basic MOSFET amplifier circuits are introduced, as well as the low and high frequency responses. Chapter 5 covers differential amplifiers, which are widely used in analog integrated circuits. In chapter 6, important operational amplifier circuits are introduced, where feedback effects play a very important role. Besides being a primer in microelectronics, this book can also be used as a study guide to supplement the learning of this topic.

I would like to express my thanks to the people whose contributions made the publishing of this book possible. First, I would like to thank my agent at IOP, Jeanine Burke, who extended the invitation to me to write this book and answered many questions. Second, I would like to thank Stephanie Howard, a former student who took this course with me, who did an excellent job in editing the manuscript. Third, I would like thank the reviewers who provided many beneficial suggestions and also corrected a few errors in the manuscript. Chapters 1 and 4 were reviewed by Camden Criddle, chapters 2 and 5 were reviewed by Dakota Crisp, chapter 3 was reviewed by David Probst and chapter 6 was reviewed by Timothy Pierce. Fourth, I would like to thank my wife, Qin Zhong, for supporting me writing this book. Last but not least, I would like to thank all my former students, who took my courses on semiconductor devices and electronic circuits in the past 14 years. While preparing the lectures and interacting with my students, I have been forced to think more deeply and find better ways to explain these topics. My experience demonstrates that most students can learn microelectronics quite well if the right approach is adopted. The motivation in writing this book is to help more students overcome the challenge of this course and start a successful career related to electronics.

Yumin Zhang
Cape Girardeau, Missouri, 2014.

Bibliography

Semiconductor electronic devices

[1] Neamen D A 2011 *Semiconductor Physics and Devices: Basic Principles* 4th edn (New York: McGraw-Hill)

[2] Streetman B and Banerjee S 2014 *Solid State Electronic Devices* 7th edn (Englewood Cliffs, NJ: Prentice-Hall)

[3] Sze S M and Ming-Kwei L 2012 *Semiconductor Devices: Physics and Technology* 3rd edn (New York: Wiley)

[4] Tsividis Y and McAndrew C 2010 *Operation and Modeling of the MOS Transistor* 3rd edn (Oxford: Oxford University Press)

[5] Hu C C 2009 *Modern Semiconductor Devices for Integrated Circuits* (Englewood Cliffs, NJ: Prentice-Hall)

[6] Fossum J G and Trivedi V P 2013 *Fundamentals of Ultra-Thin-Body MOSFETs and FinFETs* (Cambridge: Cambridge University Press)

Microelectronic circuits

[1] Sedra A S and Smith K C 2009 *Microelectronic Circuits* 6th edn (Oxford: Oxford University Press)

[2] Neamen D 2009 *Microelectronics Circuit Analysis and Design* 4th edn (New York: McGraw-Hill)

[3] Jaeger R and Blalock T 2010 *Microelectronic Circuit Design* 4th edn (New York: McGraw-Hill)

[4] Razavi B 2013 *Fundamentals of Microelectronics* 2nd edn (New York: Wiley)

[5] Floyd T L and Buchla D M 1998 *Basic Operational Amplifiers and Linear Integrated Circuits* 2nd edn (Englewood Cliffs, NJ: Prentice-Hall)

[6] Sergio F 2014 *Design With Operational Amplifiers and Analog Integrated Circuits* 4th edn (New York: McGraw-Hill)

[7] Young P H 2003 *Electronic Communication Techniques* 5th edn (Englewood Cliffs, NJ: Prentice-Hall)

[8] Lee T H 2003 *The Design of CMOS Radio-Frequency Integrated Circuits* 2nd edn (Cambridge: Cambridge University Press)

[9] Carusone T C, Johns D A and Martin K W 2011 *Analog Integrated Circuit Design* 2nd edn (New York: Wiley)

[10] Baker R J 2010 *CMOS Circuit Design, Layout, and Simulation* 3rd edn (New York, NY: Wiley)

[11] Gray P R, Hurst P J, Lewis S H and Meyer R G 2009 *Analysis and Design of Analog Integrated Circuits* 5th edn (New York: Wiley)

[12] Allen P E and Holberg D R 2002 *CMOS Analog Circuit Design* 2nd edn (Oxford: Oxford University Press)

Author biography

Yumin Zhang

 Yumin Zhang is an associate professor in the Department of Physics and Engineering Physics, Southeast Missouri State University. His academic career started in China; in 1989 he obtained a master's degree in physics from Zhejiang University and then was employed as technical staff in the Institute of Semiconductors, Chinese Academy of Sciences. After receiving a PhD degree in electrical engineering from the University of Minnesota in 2000, he started to work as a faculty member at the University of Wisconsin–Platteville and then at Oklahoma State University–Stillwater. His research fields include semiconductor devices and electronic circuits. Since joining Southeast Missouri State University in 2007, he has also investigated in the field of Physics and Engineering Education. In addition, he is very interested in teaching Chinese to non-native speakers and has written a book on this topic: *Roots and Branches: a Systematic Way of Learning Chinese Characters.*

The Tao of Microelectronics

Yumin Zhang

Chapter 1

Overview

Electronic devices are the basic building blocks of any electronic circuit; therefore, it is imperative to have an intuitive understanding of their behavior. In this chapter they are introduced in a coarse grained way—their basic features are highlighted and details are ignored. One of the major challenges for beginners is the 'double-faced' feature of the circuit elements; i.e., their characteristics in DC and AC are very different. In addition, the basic principles of rectifier and amplifier circuits are also introduced.

1.1 Basic circuit elements

If one has the experience of playing stringed musical instruments, one is well aware of the importance of tuning the instrument before playing it. It is the same with an electronic circuit: one needs to do the DC analysis first before investigating its AC behavior. DC stands for *direct current*, which indicates that current flows only in one direction. However, in most situations in microelectronics, DC refers to constant voltage or current. In contrast, AC stands for *alternating current*, which means that current flows back and forth, just like the vibration of a string in a musical instrument. Unfortunately, most circuit elements behave in quite different ways in DC and AC modes, and the only exception is the resistor. Before investigating the active electronic devices, we can first study the basic circuit elements.

1.1.1 Voltage and current sources

For linear circuits, the superposition principle applies. In many microelectronics textbooks, people adopt the following convention: DC parameters are specified by upper-case variable names with upper-case subscripts, such as V_A and I_B; on the other hand, AC variables are expressed with lower-case variable names with lower-case subscripts, such as v_a and i_b; the superposition of DC and AC signals is denoted in a

1-1   © Morgan & Claypool Publishers 2014

hybrid way, i.e. lower-case variable names with upper-case subscripts, such as v_A and i_B. With the superposition principle, they are related in the following formulae:

$$v_A(t) = V_A + v_a(t)$$
$$i_B(t) = I_B + i_b(t) \tag{1.1}$$

A DC or AC source can be viewed as a special case of the general DC–AC pair where the other component vanishes. If a voltage source component vanishes, it is equivalent to a short circuit, which guarantees the voltage equal to zero. For example, a good DC voltage source, such as a battery, generates no AC signal, so its AC behavior is just like a short circuit. As voltage sources are often directly connected to the ground, the effect of such a short circuit is equivalent to grounding the node connected to the voltage source. On the other hand, if a current source component vanishes, it is equivalent to an open circuit, since the current is equal to zero there. The current sources are easier to handle, as an open circuit often means breaking away a branch of a circuit. Therefore, the behavior of these four different types of source can be summarized as in table 1.1.

Table 1.1. Behavior of sources.

Type of source	DC behavior	AC behavior
DC V source	V source	short circuit
DC I source	I source	open circuit
AC V source	short circuit	V source
AC I source	open circuit	I source

1.1.2 Capacitors and inductors

Besides the sources, capacitors and inductors also have interesting DC–AC behaviors. For example, the DC behavior of a capacitor is the same as that of an open circuit, and that of an ideal inductor can be treated as a short circuit. In addition, in a relatively high frequency domain, the AC behavior of a large capacitor can be regarded as a short circuit, and that of a large inductor as an open circuit. These behaviors are summarized in table 1.2.

Table 1.2. Behavior of large capacitors and inductors.

Type of device	DC behavior	AC behavior
Large capacitor	open circuit	short circuit
Large inductor	short circuit	open circuit

In practice, the DC behavior of most capacitors can be considered as open circuit, since the conductivity of the dielectric materials in the capacitors is very low. However, real inductors often have noticeable parasitic resistance, as they are made from coils of thin copper wires. For example, a discrete 0.1 mH inductor can have 0.2 Ω parasitic resistance, and this value can rise to several ohms for larger inductors.

One of the major applications of capacitors in discrete analog circuits is to separate DC and AC signals; i.e., they allow AC signals to pass through but block the DC current. On the other hand, inductors are used in resonance circuits or for RF choke. The formulae of their impedance are needed in the design process:

$$Z_{\mathrm{C}} = \frac{1}{j\omega C}, \qquad Z_{\mathrm{L}} = j\omega L. \qquad (1.2)$$

In table 1.2 *short circuit* is an approximation of very low impedance and *open circuit* is an indication of very high impedance. At very high frequency, based on equation (1.2), even moderate sized capacitors and inductors can have these simplified behaviors.

1.1.3 Example

As an example, we will separate a common-base (CB) bipolar junction transistor (BJT) amplifier into its DC and AC equivalent circuits, which are shown in figure 1.1. If you have not been exposed to transistors before, don't worry about it at all: the emphasis of this example is on demonstrating the different behaviors of capacitors and various sources we have just discussed.

- **Capacitors**
  There are two capacitors (**C1** and **C2**) in this circuit, and they are equivalent to *open circuit* in DC, so only the central column remains in the DC equivalent circuit (*b*). On the other hand, they are equivalent to *short circuit* in AC, so they disappear from the AC equivalent circuit (*c*).

- **DC sources**
  In addition, there are two DC voltage sources (**V$_{\mathbf{CC}}$** and **V$_{\mathbf{EE}}$**) and one DC current source (**I$_{\mathbf{O}}$**); all of them are intact in the DC equivalent circuit (*b*), but they disappear from the AC circuit (*c*). As the AC equivalent of a DC

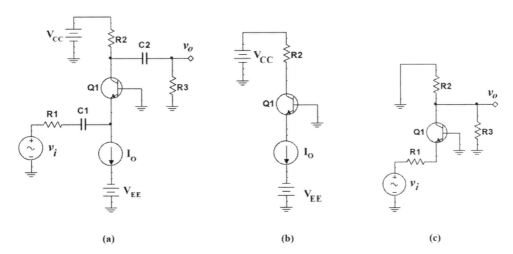

(a)   (b)   (c)

Figure 1.1. (*a*) Complete CB amplifier; (*b*) DC equivalent circuit; (*c*) AC equivalent circuit.

current source is *open circuit*, the bottom part of circuit (*a*) is removed from circuit (*c*). In addition, the AC equivalent of a DC voltage source is *short circuit*, so resistor **R2** is directly grounded at the top in circuit (*c*).

- **AC source**
  The input signal v_i is an AC voltage source, so it remains intact in circuit (*c*). Its DC equivalent is short circuit, but it is isolated from the center column by capacitor **C1** and removed from the DC circuit (*b*).

1.2 Diode—rectifier

When winter comes, tire pressures will drop, so people need to pump more air into the tires of their vehicles. Such an operation relies on a very important device, the single-sided valve, which allows air to flow in only one direction. Diodes are a family of electronic devices that play the same role—they allow current to flow in only one direction. A diode and its circuit symbol are shown in figure 1.2. The circuit symbol is very descriptive: the arrowhead indicates the permissible current direction, and the bar indicates the location of the strip printed on real diodes.

(a)   (b)

Figure 1.2. (*a*) Diode; (*b*) its circuit symbol.

The DC behavior of ideal diodes is summarized in table 1.3. When current tries to flow in the direction indicated by the arrowhead, a diode behaves like a *short circuit*. However, if current tries to flow in the opposite direction, it behaves like an *open circuit*. For real diodes, there is a voltage drop in the forward biased situation.

Table 1.3. DC behavior of ideal diodes.

Bias condition	Behavior
Forward bias	short circuit
Reverse bias	open circuit

One of its important applications is a rectifier, which can convert AC power into DC power. For example, as smart phones are very power hungry, people need to recharge the battery regularly. However, the power source from the wall socket is an AC voltage source, and it needs to be converted into a DC voltage source by an AC/DC adaptor. Although most DC voltage/current discussed in microelectronics refers to constant voltage/current, its definition is much broader. For example, a positive pulse train signal, as shown in figure 1.3, also belongs to the DC category.

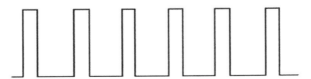

Figure 1.3. Positive pulse train signal.

If other factors are not taken into account, converting an AC signal into a DC signal is straightforward—just placing a diode on the signal path. Figure 1.4(*a*) shows an example of such a rectifier circuit, and figure 1.4(*b*) shows the simulation result of this circuit. The waveform at the top is the input signal (v_i) with an amplitude of 100 V, and the bottom waveform is the output signal (v_o). In order to show both of them together, the output waveform is shifted downwards by 110 V. Comparison of the two waveforms indicates that the diode rejects the negative half of the input signal and keeps only the positive half. This circuit is termed a *half-wave* rectifier, as the current flows only during 50% of the duty cycle. The drawback is obvious: the power conversion efficiency is cut by at least half. This problem can be overcome by using a *full-wave* rectifier circuit with four diodes.

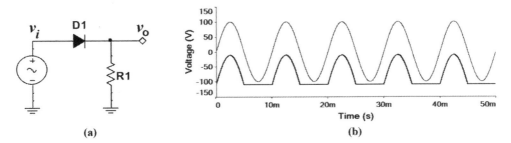

Figure 1.4. (*a*) Simple rectifier circuit; (*b*) simulated input/output waveforms.

The battery recharging process cannot be simulated easily; instead, we can replace the battery with a capacitor. Figure 1.5(*a*) shows such a circuit, and the simulated waveforms are shown in figure 1.5(*b*). The sine wave is the signal from the voltage source, and the thicker curve is the voltage across the capacitor. The simulation result indicates that the recharging is more efficient at the beginning, when the voltage across the capacitor is low. When this voltage rises gradually, the

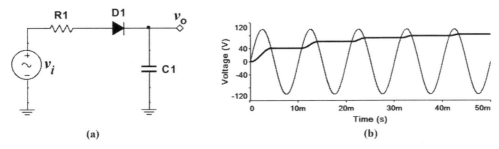

Figure 1.5. (*a*) Capacitor recharging circuit; (*b*) simulated waveform.

charging current gets weaker and the efficiency of recharging reduces progressively. However, the behavior of a battery is very different from a capacitor, and the voltage does not change much in the recharging process.

1.3 Transistor—varistor

Unlike most passive circuit elements, a transistor has three terminals. The origin of the name *transistor* is from the concatenation of two words: *trans*conductance + var*istor*. The meaning of transconductance will be discussed in the next section; now, we will concentrate on the aspect of a varistor—a resistor with a variable value. From this point of view, current will flow through two of the three terminals of a transistor, just like a resistor, while the third terminal is used for controlling the resistance. Actually, this is a pretty good description of a transistor in switching mode.

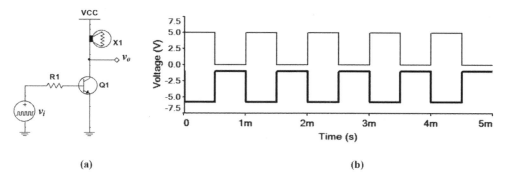

(a)   (b)

Figure 1.6. (*a*) BJT switching circuit; (*b*) input and output waveforms.

Figure 1.6 is a simple switching circuit implemented by an *npn*-type BJT; the terminal on the left (*base*) of the BJT controls the resistance between the other two terminals in the vertical direction. The V_{CC} at the top is a shorthand way to represent a DC voltage source, which is 5 V in this circuit. Below V_{CC} is a light bulb (**X1**), which is essentially a resistor (R_X). If the transistor is considered as another resistor (R_T) between the two terminals on the vertical path, these two devices form a voltage divider circuit. In this case, the voltage at the node between the light bulb and the transistor can be expressed as

$$V_O = \frac{R_T}{R_X + R_T} V_{CC} \tag{1.3}$$

The source signal (**V1**) on the left generates a positive square wave between 0 and 5 V, which is shown at the top in figure 1.6(*b*). The output waveform shown at the bottom is also a square waveform between 0 and 5 V, but it is shifted downwards by 6 V so that the overlapping between these two waveforms can be avoided. Comparison of these two waveforms shows that the high and low levels are inverted—the output is high when the input is low and vice versa. From the voltage divider formula above, it can be inferred that the resistance of the transistor (R_T) is very low when the base voltage is high, and vice versa. There is a partner for an *npn*-BJT, a *pnp*-BJT, where the *n*-type and *p*-type doping regions in the device structure are swapped. In a switching circuit, the emitters of a *pnp*-BJT are connected to V_{CC}. Figuratively speaking, the *pnp*-BJTs are hanging upside down from the ceiling in these circuits. The behaviors of these two types of BJT are summarized in table 1.4.

Table 1.4. Switching behavior of BJTs.

Base voltage	*npn*-BJT	*pnp*-BJT
High	low resistance	high resistance
Low	high resistance	low resistance

One might wonder why we need to put a transistor there. Suppose we directly connect the light bulb to the voltage source **V1**: the same square waveform can be achieved, if the inversion of voltage level is irrelevant. One example requiring such a switching device is a thermostat that regulates room temperature. As we know, the sensor and control unit on the wall is powered by one or two AA batteries operating with a current level of milliamperes, but it can turn on/off an air conditioning unit that runs at a current level of hundreds of amperes. Therefore, one of the applications of BJTs is to control a large current with a small current.

Actually, MOSFETs can do an even better job than BJTs, since they do not need any input current at all. In MOSFETs the conduction channel is insulated from the control terminal (*gate*) with a layer of silicon dioxide, and the voltage signal at the gate generates an electric field in the oxide layer, which in turn controls the resistance of the conducting channel. The first part of the name, MOS, represents the material structure of the device, metal–oxide–silicon; and the second part of the name, FET, indicates the operating principle: field effect transistor.

The MOSFET is the dominant device in the microelectronic industry, and it is the workhorse in processors and memory units of most electronic devices, such as computers and cell phones. In the field of digital electronic systems the basic building blocks are logic gates, and the simplest gate is an *inverter*; its symbol and transistor level circuit diagram are shown in figure 1.7.

Inside the inverter circuit shown in figure 1.7(*b*) there are two different kinds of MOSFET: the bottom one is called an *n*-type MOSFET and the top one is called a *p*-type MOSFET. Circuits with both of these two types of MOSFET are described as CMOS circuits—complimentary MOSFET circuits. The relationship of channel resistance to gate voltage is very similar to that of BJTs, and is summarized in table 1.5. The channel resistances of the two MOSFETs also form a voltage divider circuit. Following the behavior described in table 1.5, it can be found that the input/output characteristics of an inverter are identical to the waveform in figure 1.6—the output voltage level is inverted from the input voltage level.

The symbol of the inverter is also very interesting. First, the bubble indicates the inversion of the logic level. Second, the arrowhead symbol shows the transmission

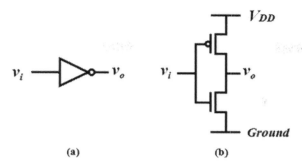

Figure 1.7. (*a*) Inverter symbol; (*b*) CMOS circuit of inverter.

Table 1.5. Switching behavior of MOSFETs.

Gate voltage	n-MOSFET	p-MOSFET
High	low resistance	high resistance
Low	high resistance	low resistance

direction of the signal. In addition, there is a hidden meaning of this triangular shape, which is the symbol of an amplifier. As we discussed above, one of the applications of a transistor is controlling a large current with a weak signal. This also applies to an inverter. For example, if the input signal is at logic low, the *p*-type MOSFET at the top of the inverter will be at its low resistance state, which allows a strong current from V_{DD} to flow through it. Furthermore, inverters can also rectify waveforms: let a deformed square wave go through two inverters in series; the deformation will be eliminated at the output. However, this causes a little delay, which needs to be taken into account in high speed digital systems.

1.4 Transistor—transconductance

As mentioned in the previous section, *transistor = trans*conductance + var*istor*. For digital electronic circuits, transistors work in the *switching mode*, and can be described as *varistors*. On the other hand, for analog electronic circuits, transistors operate in *swing mode*, and the key parameter is the transconductance. In a simplified picture, a transistor in this mode is a voltage controlled current source; the ratio of the change in the current over the variation of the control voltage is defined as the transconductance: $g_m = \Delta i_O / \Delta v_I$.

The circuits in figure 1.8 are simple amplifiers implemented with a BJT and a MOSFET, respectively. First, the objective of these circuits is to amplify the input AC signal (v_i), and the ratio between output and input AC voltage signals is called the gain of the amplifier: $A_V = v_o / v_i$. Second, one needs to analyze the DC circuit first, which is equivalent to tuning stringed musical instruments.

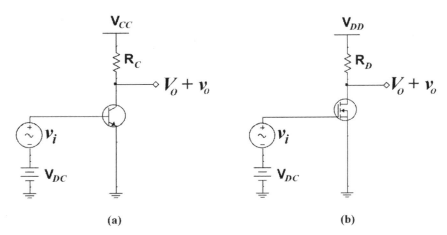

Figure 1.8. Simple amplifier circuit implemented with (*a*) BJT and (*b*) MOSFET.

The ideal DC voltage of the output node is in the middle between V_{CC} or V_{DD} and ground, so that the distortion of the output waveform can be minimized. An analogy of setting this quiescent point is doing a chin-up exercise on a horizontal bar inside a room with a rather low ceiling. If the bar is mounted too high, the head will bump the ceiling when pulled up. On the other hand, the level of the bar should not be too low either, as no one wants to bend their knees.

With Ohm's law, the output voltage in the BJT amplifier circuit can be found easily:

$$V_O + v_o = V_{CC} - (I_C + i_c)R_C. \tag{1.4}$$

Singling out the AC component, we have this simple relationship:

$$v_o = -i_c R_C. \tag{1.5}$$

As we know, the current flowing through a transistor is controlled by the voltage signal on the third terminal. If the input AC voltage signal is very weak, the relationship between these two parameters can be approximated by the following linear relationship, where g_m stands for the transconductance:

$$i_c = g_m v_i. \tag{1.6}$$

Combining the previous two equations together, the output and input signals are related with a simple equation and the voltage gain can be found:

$$v_o = -\left(g_m R_C\right)v_i \rightarrow A_V = \frac{v_o}{v_i} = -g_m R_C. \tag{1.7}$$

Beginners are often bothered by the negative sign of the gain; actually, it is harmless. For a sine wave, a negative sign is the same as a 180° phase shift. Therefore, the figure of merit is the absolute value of the voltage gain. Figure 1.9 shows the input (lower) and output (upper) waveforms of the BJT amplifier circuit, where this phase shift is shown clearly. When the input signal is at the peak, the current going through the transistor and resistor is also at the peak, and then the voltage across the resistor

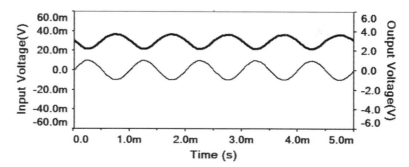

Figure 1.9. Input and output signals of BJT amplifier circuit.

is at its maximum, which pushes the voltage at the output node to the minimum. The scales of the input and output waveforms are shown on the left and right vertical axes, respectively. The amplitude of the input signal is 10 mV, and that of the output waveform is 740 mV, so the magnitude of the voltage gain is 74 V/V.

Zooming out from these details, we can see that an amplifier needs two components: the first one can convert a voltage signal into a current signal, and its capability is described by its transconductance; the second one can convert a current signal back to a voltage signal, and its effectiveness is determined by its AC resistance. In amplifier circuits, transistors play the role of the first component, and a resistor seems to be the ideal candidate for the second component. We will see later in this book that a transistor is much better than a resistor in integrated circuits (ICs). In addition, an inductor can also replace a resistor in communication circuits working at very high frequencies.

If a loading device, such as a speaker, is connected to the output node, the power of the amplified signal can be exported. Where does this power come from? One candidate is the transistor, as people often say that the behavior of transistors is just like a current source. However, this is not correct; instead, the exported AC power is from the DC power source (V_{CC} or V_{DD}). A transistor can be considered as a voltage-controlled current source, but it is not a real current source and it cannot deliver power. Actually, transistors consume power, and this is the most critical issue for power amplifiers. If a transistor consumes too much power, the efficiency of the amplifier circuit is very low. Furthermore, the large amount of heat generated can also cause irreversible damage to the transistor.

1.5 Generic amplifier

The most important parameter of an amplifier is its gain; however, there are other parameters one needs to take into account in the design process. Frequency response is an important aspect of amplifiers, which can be divided into two categories: wide band amplifiers and tuned amplifiers. In many communication electronic circuits, the gain is required to peak around a single frequency and then

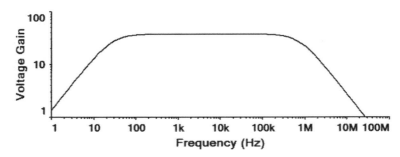

Figure 1.10. Typical frequency response of amplifiers.

drop off rapidly away from it. Such tuned amplifiers will not be discussed in detail in this book.

For wide band amplifiers, the gain is required to be constant in a relatively wide frequency range, and then rolls off beyond a lower and an upper cutoff frequency, f_L and f_H, respectively. Some amplifiers do not have the lower cutoff frequency, such as operational amplifiers (op-amps), and they can work well even with DC. However, any amplifier has an upper frequency limit, which is essentially equal to the bandwidth, though its rigorous definition is the difference between the high and low cutoff frequencies. A typical frequency response of wide band amplifiers is shown in figure 1.10, where the frequency range with a constant gain is called the *midband*. Within this region, the AC effects of all capacitors and inductors can be neglected, so the circuit analysis is much simpler.

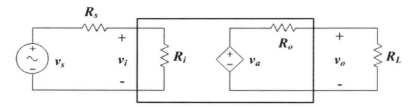

Figure 1.11. Block diagram of a generic voltage amplifier.

Figure 1.11 shows a generic amplifier circuit in the midband, where v_i and v_o are the AC voltages measured at the input and output ports of the core amplifier specified in the block, respectively. In a complete amplifier circuit, there are three sections: signal source (v_s and R_s), core amplifier, and load (R_L). With a different source and load, the gain of the whole circuit could be very different. The voltage-controlled voltage source is described by $v_a = A_{VO}v_i$, and the gain of the overall circuit can be expressed as

$$A_V = \frac{v_o}{v_s} = \frac{v_i}{v_s}\frac{v_a}{v_i}\frac{v_o}{v_a} = \frac{R_i}{R_s + R_i}A_{VO}\frac{R_L}{R_o + R_L}. \tag{1.8}$$

The first and last factors are derived from the two *voltage-divider* circuits interfacing the amplifier, and A_{VO} is the internal gain of the amplifier, which can be reached in the ideal case when $R_s = 0$ and $R_L = \infty$. However, if the gain of the circuit is expected to approach the internal gain even with finite signal and load resistances, then the input resistance R_i needs to be very high and the output resistance R_o is required to be very low. In summary, a generic amplifier in the midband has three key parameters: internal gain, input resistance, and output resistance. A good amplifier should have high internal gain, high input resistance, and low output resistance.

To obtain an intuitive understanding of such a situation, an analogy with the example of income and purchasing power is very helpful. The internal gain can be considered the nominal salary, the first factor indicates that one needs to pay income tax, and the last factor is equivalent to the deduction due to sales tax, so the overall gain is the real purchasing power. Therefore, even with a very high internal gain A_{VO}, if the input resistance is rather low and the output resistance is pretty high, then it is not a good amplifier. For example, if $R_i \approx R_s$ and $R_O \approx R_L$, which is equivalent to 50% income tax and 100% sales tax, then the purchasing power is reduced to one quarter of the nominal salary, $A_V \approx 0.25 A_{VO}$.

Usually the internal resistance of a signal source is 50 Ω, which does not cause any challenge to the input resistance of an amplifier. Furthermore, the input resistance at the gate of MOSFETs approaches infinity. On the other hand, the load resistance can be pretty low. For example, the load resistance for many speakers is 8 Ω, which is a challenge for most amplifiers without an output stage. In addition, multiple-stage amplifiers are used in many applications; in that case, the output resistance of the previous stage is equivalent to R_s, which is usually not negligible. Therefore, the overall gain of a multiple-stage amplifier is usually lower than the product of the gains at every stage.

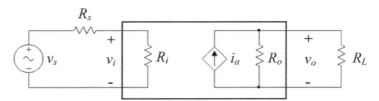

Figure 1.12. Block diagram of a generic transconductance amplifier.

Another generic amplifier configuration is shown in figure 1.12; it is pretty close to a small signal model of a BJT, which will be covered in chapter 3. The voltage-controlled current source is determined by $i_a = G_m v_i$, where G_m is the generalized transconductance. With Thévenin's theorem, this circuit can be converted into the one shown in figure 1.11, and the internal gain can be found easily: $A_{VO} = G_m R_o$. In addition, the output resistance is still R_o. Here one can see a trade-off in design: a higher R_o can boost the internal voltage gain, but it causes trouble when the load resistance is rather low. The solution to this problem is the adoption of a multiple-stage amplifier configuration with one stage achieving a high gain in voltage and another stage with a high gain in current.

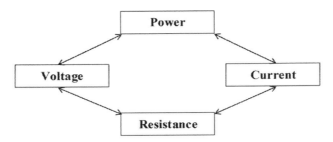

Figure 1.13. Relationship between voltage, current, resistance, and power.

The issue of output resistance is closely related to output current or power, as well as the voltage. Actually, these four parameters are closely related, forming the diamond network shown in figure 1.13. For amplifier circuits with transistors, especially for ICs, resistance is not only associated with real resistors. Instead, it often just reflects the ratio of voltage over current in AC. Therefore, if a circuit can export a large current, its output resistance becomes very low and the output power is rather high.

The Tao of Microelectronics

Yumin Zhang

Chapter 2

pn-junction diode

Before investigating the behaviors of semiconductor devices, we need a brief review of semiconductor physics. There are a few basic concepts to be introduced, such as energy bands, drift and diffusion currents, etc. Most semiconductor devices involve regions with two different kinds of doping, so a *pn*-junction is present in all of them. A diode is a simple device, which can be implemented by a *pn*-junction, as well as other hetero-junction structures.

2.1 Energy band

With knowledge of atomic physics, we know that the energy of electrons in atoms is not a continuous function; instead, there are discrete levels. The spatial distributions of electrons at different energy levels are figuratively called orbitals—just like satellites moving around the Earth at different heights. Although this picture is very intriguing, it is not accurate, and a rigorous description needs knowledge of quantum mechanics. The value of each energy level is a strong function of the atomic number, i.e. the number of protons, which is the identifier of different elements. However, the structure of these energy levels is the same for all atoms: 1s, 2s, 2p, 3s, 3p, 3d, etc. In addition, each orbital can only be occupied by a limited number of electrons: for example, two electrons in an s orbital, six electrons in a p orbital, etc. Therefore, atoms with a large number of electrons can populate higher energy levels. The onion model is a simplified description of the spatial distribution of these electrons: these at the lower orbitals stay in layers close to the nucleus and those at the higher energy levels are located in the outer layers. The electrons at the outermost layer are called *valence* electrons.

When two atoms are put very close together, the electrons in the lower orbitals are not affected. However, the orbitals of the valence electrons can overlap, allowing these electrons to travel between the two atoms. In this case, each of the energy levels of

2-1 © Morgan & Claypool Publishers 2014

these shared electrons will split into two energy levels. When a stable bond is formed, electrons will occupy the lower energy level and leave the higher energy level empty. In a crystal, such overlapping orbitals occur throughout the whole material. As a result, each atomic energy level occupied by these valence electrons expands into a wide energy band, which is composed of many discrete energy levels placed close together. Although the real situation is much more complicated, this simplified picture gives us an intuitive understanding of the energy band structure of crystals.

Between atomic energy levels, there are wide energy gaps. Similarly, when the energy levels of the valence electrons expand into energy bands, usually there is also a gap between the two bands. Many important properties of materials are determined by this energy gap, which is simply called the *bandgap*. For example, if the bandgap is closed up, it is a good conductor, such as a metal. In contrast, if the bandgap is very wide, it is an insulator. Semiconductors fall between these two extreme situations. Additionally, the bandgap also determines many optical properties of the material. For example, materials with wide bandgaps are transparent.

Silicon is the most important semiconductor in microelectronics. Its atomic number is 14, which means that there are 14 protons in the nucleus and 14 electrons moving around it. Ten electrons populate the inner orbitals ($1s^2$, $2s^2$, $2p^6$), and the remaining four ($3s^2$, $3p^2$) are valence electrons. The silicon crystal used in microelectronics has a diamond lattice structure, where each atom is bonded to four others in a symmetric tetrahedral form. Due to this symmetry, the orbitals of the four valence electrons become *hybridized* (reorganized); i.e., the original atomic s and p orbitals disappear and a new form of orbitals emerges. If one has trouble understanding this process, it may be of some help to compare this process to a personal relationship. Before two people get married, each of them has an individual life style. However, once married and living together, they have to make some compromises and try to accommodate their partner's life style. After a short (hopefully) break-in period, a shared, new life style starts to emerge in their family life. If a child is born into the family, the couple has to make adjustments again: for example, reducing the frequency of dining out and giving up weekend movies. Failure in foregoing one's old life style and adopting a new one often causes various family issues or even divorce. Similarly, the formation of crystals can be viewed as individuals joining the military, where people must behave in a collective manner.

In the course of the hybridization process, something remains unchanged. For example, in an isolated silicon atom, two valence electrons occupy the 3s orbitals and another two electrons occupy part of the 3p orbitals, leaving four openings in the 3p orbitals. In other words, 3s3p orbitals can accommodate eight electrons and silicon atoms only use half of that space. After the silicon crystal is formed, these hybridized orbitals are divided into two groups with equal capacity, and they are separated by a bandgap. As a result, the group with lower energy (*valence band*) is fully occupied by electrons and the group with higher energy (*conduction band*) is essentially empty. This feature is present in most intrinsic (without impurity) semiconductor materials, such as Ge, InP, GaAs, and GaN.

Figure 2.1 illustrates the transition from discrete atomic levels (on the right) to quasi-continuous energy bands (on the left) as the interatomic distance (horizontal axis) shrinks. Initially, as the atoms become closer, the energy levels broaden and

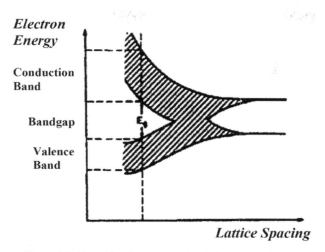

Figure 2.1. Transition from energy levels into energy bands.

the gap reduces; at a certain distance, the gap disappears. However, as the atoms continue getting closer, the gap reappears and increases with further shrinking of the interatomic distance. One way to change this distance is by applying intense pressure. This approach is used in advanced research laboratories in the investigation of the band structure of various materials. Another way to achieve this is by changing temperature; it is well known that materials expand with increasing temperature, and such expansion is the result of increased interatomic distance. This effect can be demonstrated simply by putting an LED into liquid nitrogen briefly (its boiling point is at 77 K). A blue-shift of the emitted light can be observed, which is an indication of increased bandgap. Table 2.1 lists the bandgap energy of several widely used semiconductors and an insulator at room temperature and atmospheric pressure.

Table 2.1. Bandgap energy.

Semiconductor	Bandgap (eV)
Ge	0.67
Si	1.12
InP	1.35
GaAs	1.42
GaN	3.4
SiO_2	9

At first sight, all the intrinsic semiconductors should be insulators. With energy levels in molecular bonds as an analogy, there should be no electrons in the conduction band, and the valence band should be fully occupied. This situation can be imagined as a two-story school building with the classrooms on the first floor fully occupied and those on the second floor completely empty. Although the students are allowed to change seats, they cannot do so as there is no open seat available on their level.

However, if some students are allowed to go upstairs, then opportunities to move are created. This is the way semiconductors conduct electricity.

A few factors determine the probability of electrons in the valence band jumping up to the conduction band. First, the bandgap energy plays a dominant role: if the bandgap is very large, such as in quartz (SiO_2 crystal), few electrons are energetic enough to make the transition. As a result, SiO_2 is a very good insulator for electronic devices, such as MOSFETs. Second, increasing temperature enables more electrons to gain enough energy to jump up to the conduction band. These excited electrons and the space left behind in the valence band provide more opportunities for motion. Therefore, unlike metals, semiconductors' conductivity has a positive temperature coefficient, i.e. they become more conductive at higher temperatures. While temperature may also slightly influence the bandgap, this effect is rather weak and considered negligible.

2.2 Drift current

An intrinsic semiconductor is merely a theoretical model, and various impurity atoms are always present in real semiconductor materials. In the early history of the semiconductor industry, purification was an important advancement. If the purity of silicon is higher than 99.9%, it can be used for solar cells. However, for the fabrication of ICs the requirement is much higher, often with the impurity level less than 10^{-9}. In the field of semiconductors, the preferred unit of atomic concentration is the number of atoms per cubic centimeter (cm^{-3}). For example, the concentration of silicon atoms is about $5 \times 10^{22}\,cm^{-3}$ and the impurity concentration is required to be below $10^{14}\,cm^{-3}$.

When foreign impurities replace the host atoms, they create energy levels inside the bandgap. Some of them create energy levels located in the middle of the bandgap, which are called deep impurities. These include most metal elements, such as Au, Cu, Mn, Cr, and Fe, and their energy levels are shown in figure 2.2. These impurities are harmful to the operation of most semiconductor devices and they need to be eliminated. On the other hand, some impurity atoms create energy levels very close to the edge of the valence or conduction band. They are called shallow impurities and are very useful in manipulating the properties of semiconductors.

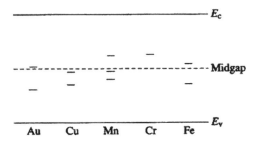

Figure 2.2. Energy levels of some deep impurity atoms.

These shallow impurity atoms are intentionally doped into semiconductors, and this is an important step in the fabrication of ICs.

The shallow impurities are divided into two different categories: *n*-type impurity atoms with five valence electrons (such as As and P), and *p*-type impurity atoms with three valence electrons (such as B). When an *n*-type impurity atom is incorporated into a silicon crystal, four out of the five valence electrons will bond with the neighboring four silicon atoms, but the remaining electron is free to go into the conduction band and move about. This free moving electron is called a *carrier*, because it has the capability to transport electric charge. As electrons are negatively charged, they are called *n*-type carriers. In addition, as these free moving electrons are *donated* from these impurity atoms, so they are called *donors*. On the other hand, when a *p*-type impurity atom is doped into silicon crystal, an electron vacancy, which is called a *hole*, is generated in the valence band. As a result, the electrons nearby can move over and fill in, and this also contributes to the motion of electrons and results in conduction of current.

In order to understand *p*-type doping better, the analogy of a classroom with students can be used again. Suppose there is a huge classroom filled with students but with only one open seat. As the lecture proceeds, a student sitting next to this open seat occupies it. After a while, his neighbor takes the original seat of the first student. The instructor on the stage cannot see which student is hopping seats, and his perception is that the open seat is moving about in the classroom. Similarly, a hole in a semiconductor is considered as a carrier with positive charge moving in the valence band, which also has mass and other properties of a quasi-particle. Therefore, holes are considered *p*-type carriers, and impurity atoms of this type are called *acceptors*.

One important property of carriers is their mobility, which is defined as the ratio between drift velocity and electric field. Equation (2.1) shows the simple relationship between these parameters. An analogy of skiing might be helpful: the electric field can be imagined as a mountain slope covered with snow, the velocity is the speed of the skier, and then the mobility describes the condition of slope. At first sight, this equation does not make sense; from Newton's second law, electric field is directly related to the force and thus should be proportional to acceleration instead of velocity. The classical explanation is like this: trees are grown densely on the slope and the skier bumps into these trees and falls down regularly, creating an unpleasant experience at the ski resort. Therefore, the situation is not motion with constant speed, but with many short sections of acceleration. Thus, the velocity in equation (2.1) can be understood as average velocity, which depends on the magnitude of acceleration and the free flight time between two succeeding collisions.

$$\vec{v} = \mu \vec{E}. \tag{2.1}$$

What are the physics factors contributing to these 'trees'? An obvious candidate is the impurity atoms in the semiconductor crystals, so the mobility decreases as doping concentration increases. Another culprit is temperature, which causes vibration of the lattice. Therefore, the mobility also decreases as temperature rises. In addition, the mobilities of electrons and holes are different; usually electrons have a higher mobility than holes. Table 2.2 shows the mobilities of three kinds of semiconductor material at room temperature with low doping levels. Although the mobilities are often listed as

Table 2.2. Mobilities of electrons and holes.

Semiconductor	$\mu_e(\text{cm}^2\ \text{V}^{-1}\ \text{s}^{-1})$	$\mu_h(\text{cm}^2\ \text{V}^{-1}\ \text{s}^{-1})$
Si	1350	480
Ge	3900	1900
GaAs	8500	400

positive values for both electrons and holes, they actually move in opposite directions in electric field, as holes are considered positively charged particles.

From the discussion above, one can find that mobility is not a *clean* concept with a solid physics foundation. Instead, it is just a linear coefficient of velocity versus electric field. Fortunately, in the low electric field region ($<10^3\ \text{V cm}^{-1}$), there is a simple linear relationship between these two important parameters for most semiconductor materials. However, when the electric field becomes higher, the relationship becomes nonlinear; for GaAs the mobility of electrons even starts to decrease beyond $3 \times 10^3\ \text{V cm}^{-1}$, and this negative slope is equivalent to negative resistance that can be used in generating oscillation in microwave circuits. At very high electric fields, the drift velocity will saturate; for silicon this happens at around $5 \times 10^4\ \text{V cm}^{-1}$. With the gate length of modern MOSFETs reduced to the deep submicrometer level, this critical value can be reached easily. For example, with a gate length of 40 nm and 0.5 V applied between the drain and source, the average electric field in the conducting channel is above $10^5\ \text{V cm}^{-1}$.

$$\vec{J} = \sigma \vec{E}. \tag{2.2}$$

Another simple linear relationship related to electric field is the microscopic Ohm law, which is shown in equation (2.2), where the parameter σ is the conductivity. This law is related to the more familiar macroscopic Ohm law: $I = V/R$. Assume there is a rod of uniform conductor with length L and cross-sectional area A. Its resistance can be found using $R = L/(\sigma A)$. In addition, $J = I/A$ and $E = V/L$, so everything falls into its place. In equation (2.2), the conductivity σ is defined as the ratio between current density and electric field. It is determined by two factors: carrier density and mobility. In order to understand this, we can backtrack one step and find the expression of the current density from a different perspective, which is shown in equation (2.3).

$$\vec{J} = qn\vec{v}. \tag{2.3}$$

The carriers can be imagined as vehicles carrying charges moving on a highway, and then one can stand on an overpass and count how much charge goes through per second. Suppose all the carriers are identical and each of them carries a charge q, their density is n, and they all move with the same speed of v. In this way, the current density J represents the quantity of charges carried across a unit cross-sectional area per second. Plugging equation (2.1) into equation (2.3), and then comparing it with equation (2.2), one can find the expression of conductivity: $\sigma = q\mu n$. In semiconductors, there are free moving electrons in the conduction band and holes in the valence band, and both of them contribute to the current. Therefore, the complete expression should include both types of carrier, as shown in equation (2.4).

$$\sigma = e(\mu_n n + \mu_p p). \tag{2.4}$$

In semiconductors, the concentrations of these two types of carrier are significantly different. Usually one kind of impurity is intentionally introduced, the resulting carrier is the majority, and then the other kind of carrier becomes the minority. For n-type doping, electrons are the majority carrier and holes are the minority carrier. The concentrations of these two kinds of carrier are related by the following simple equation:

$$np = n_i^2. \tag{2.5}$$

The parameter n_i in equation (2.5) is called the *intrinsic* carrier concentration. Suppose all the impurity atoms can be removed completely, and then all the conducting electrons in the conduction band have originated from the valence band. Therefore, the number of electrons in the conduction band is exactly the same as the number of holes left in the valence band. In this situation, the density of n-type or p-type carrier is defined as n_i. For example, $n_i \approx 10^{10}\,\mathrm{cm}^{-3}$ for silicon at room temperature. The real picture is very dynamic: on one hand, there are numerous electrons in the valence band and some of them can obtain enough energy to jump up to the conduction band; on the other hand, the excited electrons in the conduction band can fall down to the valence band whenever they encounter a hole. Eventually, a dynamic equilibrium is reached and n_i is the resulting concentration of electrons in the conduction band.

If shallow impurity atoms are doped into a semiconductor, almost all of them will be ionized, meaning an electron transition between the impurity level and the nearby band. For example, if arsenic atoms are introduced into a silicon crystal, the impurity energy level is 50 meV below the conduction band edge. However, the extra valence electron from the arsenic atom will not stay at its original site and will move about in the silicon crystal, as it can gain 50 meV easily from the lattice vibration and jump into a wider playground. Therefore, the carrier concentration of the majority carrier can be approximated by the doping concentration, while the concentration of the minority carrier can be calculated from equation (2.5). Actually, the large concentration of majority carriers can suppress the number of minority carriers. For example, if the doping concentration of arsenic atoms is $10^{17}\,\mathrm{cm}^{-3}$, and then the concentration of holes is lowered to around $10^3\,\mathrm{cm}^{-3}$. The huge difference in concentration makes the contribution to drift current from minority carriers negligible.

2.3 Diffusion current

When a teaspoon of juice is added to a cup of water, it will spread out quickly and eventually becomes uniformly diffused. In general, whenever there is a difference in concentration of something in a material, diffusion will occur. Such a process in liquid or gas is familiar to all of us; actually, it also happens in solid materials, though high temperature is often needed.

In the early history of the semiconductor industry, the doping process was achieved by diffusion. For example, BJTs were fabricated by triple diffusion processes. One form of diffusion uses a solid source. For example, a layer of n-type or p-type dopant is deposited on a silicon wafer first, and then it is placed into a diffusion chamber at a high temperature. These dopant atoms are activated by heat and then diffuse deeply into the silicon wafer.

$$J = -D\frac{\mathrm{d}\phi}{\mathrm{d}x}. \tag{2.6}$$

In one dimensional situations, the flux density of particles can be described by Fick's law, which is shown in equation (2.6). The diffusion coefficient D describes the *ease* with which one substance may diffuse in another substance, and it is quite high in fluids and rather low in solids. The derivative in the equation is the gradient of a certain concentration, or the slope of change in space. By definition, if $\phi(x)$ increases with x, its derivative is positive. However, the direction of diffusion is always from the high concentration region to the low concentration region, so a negative sign is needed to indicate the direction of diffusion.

Einstein investigated Brownian motion in 1905, the year he discovered special relativity. When a small particle is floating on the surface of a liquid, it will move about in a random way, caused by the collisions with the agitating liquid molecules around it. This phenomenon is called Brownian motion and it could not be well explained at that time with classical mechanics, as there are about 10^{21} collisions per second. If one put a bunch of such particles at one spot, they will spread out gradually, which is the same as the diffusion process. Einstein investigated how the density of Brownian particles will change over time, and the differential equation he set up was the following:

$$\frac{\partial \rho}{\partial t} = D\frac{\partial^2 \rho}{\partial x^2}. \tag{2.7}$$

Actually, this is often called the *continuity equation* in other situations, and it is an expression of conservation of matter. Sometimes it is easier to understand differential equations by tracing back their derivation process; for example, equation (2.7) was derived from a difference equation:

$$[\rho(x, t + \Delta t) - \rho(x, t)]\Delta A \Delta x = -[J(x + \Delta x) - J(x)]\Delta A \Delta t. \tag{2.8}$$

In this equation a common factor ΔA is added on both sides, which is the cross-sectional area. The left-hand side refers to the change of particle number in a small region with length of Δx during the time interval of Δt, and the right-hand side describes the imbalance between outlet and inlet flows during the same time interval. As the particles cannot be created or annihilated in this process, the change in their number is solely caused by the imbalance in flows across the two boundaries.

Einstein also found the solution to equation (2.7), which is shown in equation (2.9). Originally the particles are concentrated at $x = 0$ with the initial density of ρ_0; as time passes, the distribution will gradually spread out in space. Such an equation can be used to describe the diffusion doping process of semiconductors.

$$\rho(x, t) = \frac{\rho_0}{\sqrt{4\pi Dt}} \exp\left(-\frac{x^2}{4Dt}\right). \tag{2.9}$$

If diffusion were the only mechanism of migration, the world would be much simpler and very boring, as the Universe would be homogeneous. Fortunately, there are interactions between particles, which prevent diffusion from taking over the world. In cosmology, gravity pulls particles together and forms stars and planets. Electromagnetic interaction is much stronger, and it is the dominant force for the formation of molecules and condensed matter.

In semiconductors, electron concentration can also vary in different regions, and the diffusion of charged particles will result in current. The formula of such diffusion current density is very similar to Fick's law, but there is an additional factor of electric charge, which is described in equation (2.10):

$$J = -qD\frac{d\phi}{dx}. \tag{2.10}$$

For electrons, $q = -e$ and $\phi = n$; for holes, $q = e$ and $\phi = p$; where e is the (positive) magnitude of electron charge. The equations of diffusion currents for electrons and holes are presented in equation (2.11):

$$J_n = eD_n\frac{dn}{dx}, \qquad J_p = -eD_p\frac{dp}{dx}. \tag{2.11}$$

Suppose there is a silicon wafer with n-type doping by the diffusion process described above, and the dopant distribution will follow equation (2.9). What about the carrier distribution? One possibility is homogeneous distribution, as the diffusion process of carriers is much faster than that of the dopant atoms. If that is true, most donor atoms in the high concentration area will lose one electron and become positively charged, and the low concentration region will gain additional electrons and become negatively charged. Therefore, a strong electric field will build up inside the wafer, which will drive the conducting electrons back into the high concentration region. Eventually equilibrium is reached, and a weak electric field will be present in the wafer in order to counter the diffusion force. Thus, the current density for both electrons and holes should vanish. With both drift and diffusion currents taken into account, the overall current density for electrons and holes can be expressed as

$$J_n = e\mu_n nE + eD_n\frac{dn}{dx}$$
$$J_p = e\mu_p pE - eD_p\frac{dp}{dx}. \tag{2.12}$$

Detailed analysis of this problem leads to a simple relationship between diffusion coefficient and mobility, which is called the Einstein relation:

$$\frac{D_n}{\mu_n} = \frac{D_p}{\mu_p} = \frac{kT}{e} = V_T. \tag{2.13}$$

In the equation above, k stands for the Boltzmann constant (1.38×10^{-23} J K^{-1}), kT is the thermal energy, and V_T is called the thermal voltage. At room temperature, $V_T = 25.9$ mV, which will be used when analyzing diodes and BJTs. Equation (2.13) implies that the diffusion coefficient of carriers is proportional to the mobility, and also proportional to the temperature in kelvin. However, as mobility decreases with the increase of temperature, the temperature dependence is sublinear.

2.4 *pn*-junction

Although some devices can be fabricated from semiconductors with a single type of doping, such as photoconductors, most devices need both types of doping, and the interface between these two regions forms a *pn*-junction. It can be imagined, if such a junction is formed suddenly, these free moving electrons in the conduction band of the *n*-type region will diffuse into the *p*-type region, and then jump down to the valence band and fill the holes there. However, this process cannot last long, as a strong electric field will build up quickly. Before these two regions come into contact, both of them are electrically neutral. After some electrons migrate across the interface, the *n*-type region near the interface becomes positively charged, and the *p*-type region on the other side becomes negatively charged, which is shown in figure 2.3. When additional electrons try to migrate, they will experience a strong electric field pulling them back into the *n*-type region. Therefore, equilibrium is eventually achieved at the interface.

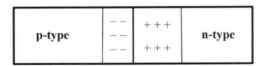

Figure 2.3. *pn*-junction.

In order to simplify analysis, clear cut boundaries are drawn between the neutral regions on both sides and the *space charge region* in the middle. As the doping concentrations of *p*-type acceptors (N_A) and *n*-type donors (N_D) are usually different on both sides, the width of space charge regions (x_p and x_n) varies accordingly, and the charge neutrality condition must be satisfied:

$$N_A x_p = N_D x_n. \tag{2.14}$$

This equation indicates that all the free moving electrons in the space charge region on the *n*-type side migrate to the *p*-type side and neutralize all the holes in the space charge region there. Therefore, there are no more carriers left in the whole space charge region, so it is also called the *depletion region*. However, this picture is oversimplified. As we know, such an abrupt change in carrier concentration is not likely to happen; instead, there is a smooth transition across the space charge region, and the majority carriers gradually become minority carriers on the other side, which is shown in figure 2.4. Because the number changes so dramatically in such a short distance, the depletion model is a good approximation in estimating the width of this region.

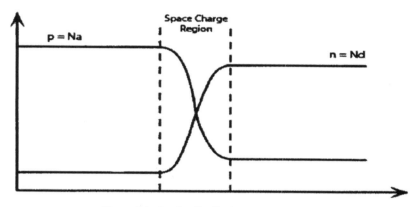

Figure 2.4. Carrier distribution in *pn*-junction.

Since there is an electric field built up in the space charge region, the edges of the conduction band and the valence band on both sides are not aligned. Figure 2.5 shows the band diagram, and the shift of the bands is determined by the leveled *Fermi energy*, which is indicated by the dashed line in the diagram. If electron concentration in the conduction band is considered similar to air density in the atmosphere, then the Fermi energy is just like sea level, so it is often called the *Fermi level*. The criterion of equilibrium is that the Fermi level is flat.

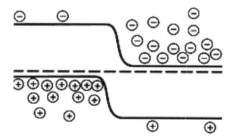

Figure 2.5. Band diagram and carrier distribution of *pn*-junction. (Courtesy of OpenStax CNX.)

As one climbs a very high mountain, the air becomes thinner. In other words, the air density is lower there. In a similar way, the probability for electrons to occupy a certain energy level depends on its 'height' from the Fermi level. For example, in the *n*-type region there are many electrons in the conduction band, so its edge is pretty close to the Fermi level. On the other hand, in the *p*-type region the concentration of electrons in the conduction band is very low, so its conduction band edge is well above the Fermi level. A good approximation to calculate the electron occupation probability at energy levels not too close to the Fermi level is the Boltzmann distribution, which is simply an exponential function:

$$p(E) = \exp\left(-\frac{E - E_F}{kT}\right). \tag{2.15}$$

Band diagrams usually just show the bandgap and the edges of the conduction and valence bands, but we need to keep in mind that electrons can stay in a wide range above the conduction band edge, just as the atmosphere extends high above the surface of the Earth.

The situation of electron distribution is a little more complicated, and the straight column of air is replaced with a bowl shaped column with its cross-sectional area increasing with the 'height'. Fortunately, it only increases fairly slowly, with a function proportional to $\sqrt{E - E_C}$. The steep exponential function of the Boltzmann distribution makes most of the electrons stay close to the conduction band edge. With this factor taken into account, electron concentration in the conduction band can be calculated in a simple way:

$$n = N_C \exp\left(-\frac{E_C - E_F}{kT}\right). \tag{2.16}$$

The parameter N_C is a constant; for silicon, it is equal to $2.8 \times 10^{19}\,\text{cm}^{-3}$ at room temperature. In most situations, the position of the Fermi level is inside the bandgap. However, if the doping concentration is extremely high, the Fermi level can move into the conduction band or the valence band. In such situations, often called *degenerate doping*, the Boltzmann distribution is no longer valid and one has to use the Fermi–Dirac distribution function instead. Let us carry out an estimation with equation (2.16): when the Fermi level is just at the conduction band edge ($E_F = E_C$), then the doping concentration is approximately $N_D \approx n = N_C \approx 10^{19}\,\text{cm}^{-3}$.

At equilibrium there is no net current; however, it is not a static situation. Instead, there is a constant exchange of electrons and holes between *n*-type and *p*-type regions. Furthermore, the generation and recombination processes between the valence band and conduction band are also very active. Figuratively speaking, electrons jump up and down and move back and forth constantly, but the overall effect is cancelled out.

2.5 Diode current and models

If two electrodes are connected to the opposite sides of a *pn*-junction, it becomes a diode. Suppose the *n*-side is grounded; a positive (*forward*) or negative (*reverse*) voltage can be applied to the p-side and the resulting *I–V* characteristic is shown in figure 2.6.

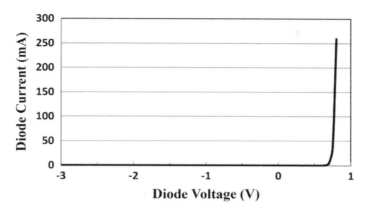

Figure 2.6. *I–V* characteristic of diode.

With reserve bias, the current becomes saturated at a very low level. For discrete diodes, the reverse saturation current is usually at $I_S \sim 10\,\text{nA}$. This value is proportional to the cross-sectional area of the *pn*-junction; in other words, a *large* diode is equivalent to many small diodes in parallel. Therefore, for *pn*-junctions in ICs, the reverse saturation current is much lower. In addition, if relatively high reversed bias voltage (~10 V) is applied, the current does increase a little. This is caused by the generation process in the widened space charge region. Furthermore, if very high reverse bias voltage (~100 V) is applied, it could break down. This breakdown voltage depends on the doping profile of the diode. If both sides of the junction are heavily doped, it becomes a Zener diode, where electrons can tunnel through the bandgap without causing damage to the device.

If an AC signal is superposed on the reverse biased DC voltage, the small signal model of a diode is simply a capacitance, which is often called *junction capacitance* (C_j). The structure of a *pn*-junction shown in figure 2.3 resembles a parallel plate capacitor: a layer of insulator sandwiched between two conductors. The value of junction capacitance can be calculated with the same formula as a parallel plate capacitor:

$$C_j = \varepsilon \frac{A}{d} = \frac{\varepsilon_r \varepsilon_o A}{x_p + x_n}. \qquad (2.17)$$

Here, the parameter d is the width of the space charge region, which can be modulated by changing the reverse bias voltage. Therefore, it becomes a voltage controlled capacitor. Such a device is called a *varactor*, a concatenation of *var*iable + cap*acitor*. Another method of concatenation gives it the name *varicap*. As we know, the resonance frequency of an RC tank circuit is given by $f_o = 1/(2\pi\sqrt{LC})$. It is pretty hard to change the value of an inductor, so the only option falls on the capacitor. In early times, radio stations were manually selected by listeners, and this was achieved by changing the overlapping areas of metal plates. Modern radios can scan the stations electronically, and a varactor can be used for this application. In addition, another important application of the varactor is in the *voltage controlled oscillator* (VCO), an important component in many communication systems.

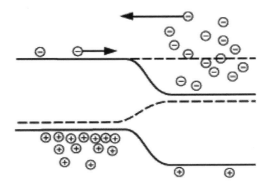

Figure 2.7. *pn*-junction with forward bias. (Courtesy of OpenStax CNX.)

When a diode is forward biased, the current will increase exponentially. Does this sound familiar? Yes: it is caused by the Boltzmann distribution of electrons in the conduction band, as well as its mirrored distribution of holes in the valence band. Figure 2.7 illustrates the imbalance in electron distribution with forward bias. It can be imagined that the same change will also happen to the holes, which is not shown in this diagram. If the analogy of atmosphere is reused, and the *p*-type region is like a very high plateau and the air above it is very thin, the *n*-type region is like an ocean, and the external bias plays the role of raising or lowering the height of the plateau quickly. With reverse bias, the height of the plateau is raised and the air will flow towards the ocean. However, since the air above the plateau is very thin, such a current is very weak. On the other hand, with forward bias, the height of the plateau is suddenly lowered considerably, and then the dense air above the ocean will flow inland. If the imbalance of the generation–recombination process in the space charge region can be ignored, the *I–V* curve can be described with a simple equation:

$$I_D = I_S\left[\exp\left(\frac{eV_D}{kT}\right) - 1\right] = I_S\left[\exp\left(\frac{V_D}{V_T}\right) - 1\right]. \qquad (2.18)$$

If the forward bias voltage is much higher than the thermal voltage V_T (25.9 mV at room temperature), then the exponential term is much greater than unity, so this formula can be further simplified:

$$I_D = I_S \exp\left(\frac{V_D}{V_T}\right). \qquad (2.19)$$

If a weak and low frequency signal is superposed on a forward DC bias voltage, its AC behavior is just like that of a resistor. There are several names for this, such as AC resistance, incremental resistance, or differential resistance. The last one refers to how to derive it from the expression of the *I–V* curve:

$$r_d = \left(\frac{dI_D}{dV_D}\right)^{-1}\bigg|_Q = \frac{V_T}{I_D}. \qquad (2.20)$$

The derivative in this equation, dI_D/dV_D, is the slope of the tangent at a certain point on the I–V curve shown in figure 2.6. This differential resistance is a strong function of DC current, which is often indicated as the Q-point of a bias circuit.

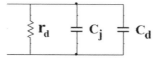

Figure 2.8. Small signal model of *pn*-junction diode with forward bias.

If the frequency of the AC signal increases, the capacitive aspect of a *pn*-junction will show up. First of all, the junction capacitance is still there, and its value is higher than the reversely biased situation, because the width of the space charge region gets smaller. In addition, as mentioned in the previous section, carrier concentration is a continuous function across the space charge region, and the applied AC voltage signal causes a variation in carrier distribution. This gives rise to another capacitance, which is called *diffusion capacitance* (C_d). In most situations, C_d is higher than C_j. A complete small signal model of a forward biased *pn*-junction is shown in figure 2.8.

IOP Concise Physics

The Tao of Microelectronics

Yumin Zhang

Chapter 3

BJT amplifier circuits

At the dawn of the 20th century, there were two fledgling industries in desperate need of amplifiers: long distance telephone systems and radio broadcasting. When the thermionic triode was invented in 1907, the dream of signal amplification was realized, though these vacuum tubes were bulky and power hungry. The bipolar junction transistor, which is much more compact and efficient, was invented 40 years later, and it brought about a wave of electronic revolution.

3.1 Vacuum tubes

From a broad perspective, a vacuum tube is very similar to a light bulb, and they share the same inventor: Thomas Edison. A carbon filament was used in early light bulbs, which gave off soot that can darken the glass enclosure. Edison put an additional plate shaped electrode into the bulb so that it could collect some of the soot and extend the lifetime of the light bulb. As an excellent experimentalist, he noticed something unexpected: current could flow between the filament and the plate electrode when it was positively biased, but there was no current when it was negatively biased. In 1883 he patented this device, and this phenomenon was called the Edison effect. Other people, such as John Fleming, further investigated this device and found that it could be used to detect radio waves. Because there are only two electrodes inside the vacuum tube, it was named a *diode*. Compared with the semiconductor diodes, the I–V characteristics are quite different: $I = KV^{3/2}$, which is much *softer* than the sharp exponential increase of the current exhibited by the semiconductor diode.

When Edison was experimenting on his vacuum tubes, the electron was not physically identified, so the current between the filament and the plate was called a *cathode ray*. In 1896, J. J. Thomson and his collaborators demonstrated that the cathode ray was the flow of a kind of negatively charged particle, which was later

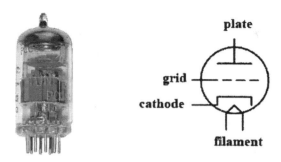

Figure 3.1. Structure of a triode vacuum tube.

named the *electron*. With this discovery the process of electron emission was further investigated, and Owen Richardson summarized the mathematical formula for the current density of this thermionic emission. The physics picture then became quite clear: in the filament there is an energy gap between the electron energy level and the vacuum energy level with freedom to travel in space, the heating of the filament provides some electrons with enough energy to jump up into the vacuum energy level, and finally the electric field between the filament and the plate attracts these electrons away and generates the current.

In 1907 the thermionic triode (figure 3.1) was invented when a third electrode with the structure of a metal mesh was inserted between the filament cathode and the cylindrical plate anode. Now the current collected by the plate could be modulated by the voltage on the third electrode, which is called the *grid*. For example, if a negative voltage is applied to the grid, the plate current will decrease. In this way, it worked like a depletion mode transistor. Figure 3.2 shows a group of typical I–V curves, with the grid voltage decreasing from left to right.

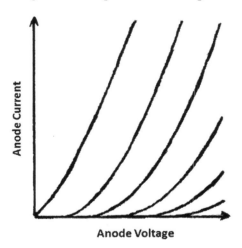

Figure 3.2. I–V characteristics of a triode vacuum tube.

Due to the structure of the grid, most electrons can penetrate it and thus the current in the grid is rather low. However, the shape of the I–V curve still looks like that of a diode; in other words, the slope is rather high and the output resistance is very low. In addition, the coupling capacitance between the grid and the plate was quite high, and the resulting large Miller effect—which will be discussed in section 3.6—was a severe problem for amplifier circuits. In order to make improvements, another mesh electrode was inserted, and it became a *tetrode*. In this way the I–V curve became flat, but kinks appeared when the plate voltage was not high enough. For further improvement, an additional mesh electrode was inserted and then it became a *pentode*, which removed these kinks in the I–V curve. Furthermore, a *hexode* with six electrodes was also developed and used in radio circuits.

The cathode of early vacuum tubes was the filament; in order to release electrons efficiently, it needed to be heated to bright incandescence. Therefore, the lifetime was rather short, typically around 100 h. Initially the research effort was focused on searching for materials with a high boiling point, such as tantalum and tungsten. As we know, the heat transfer rate in radiation is proportional to T^4, so these filaments consumed large amounts of power. Later on, materials with lower work functions (energy gaps) were discovered, such as barium and strontium oxides. These materials can release electrons efficiently at much lower temperatures, so they can be heated indirectly by a separate filament nearby. Besides, a higher vacuum level (less residue gas) inside the tube was also achieved. All these improvements extended the lifetime of the vacuum tube substantially, and made it a practical device in amplifier circuits, widely used in transcontinental telephone systems and radio receivers in airwave broadcasting.

Besides these applications in the field of analog electronics, vacuum tubes can also work in switching mode, so logic gates can be designed and implemented with them. In 1944, Tommy Flowers built a successful primitive computer with fewer than 3000 vacuum tubes in the UK, and it was used by the British to break encrypted codes from Germany. After the war, a computer named ENIAC was built with 17 468 vacuum tubes in the US, which consumed 150 kW of power. This was the first general purpose digital computer and ushered in the information age.

3.2 Introduction to the BJT

The solid state transistor was invented in 1947 in Bell Labs. Initially the BJT played the dominant role in the electronic industry, but the MOSFET took over three decades later when large scale IC technology was developed. The idea of BJT follows that of the triode vacuum tube very closely: electrons are emitted from one electrode (*emitter*) and collected from another electrode (*collector*), and there is a third electrode to modulate the current, which is called the *base* as it is located at the bottom. Figure 3.3 shows the structure of an *npn*-BJT; if the *n*-type and *p*-type regions are swapped, it will become a *pnp*-BJT. Unlike a MOSFET, a BJT is an asymmetric device, as the emitter is rather thin and heavily doped and the collector is quite wide and lightly doped. The difference in doping concentration is denoted by n^+ and n^- in the diagram.

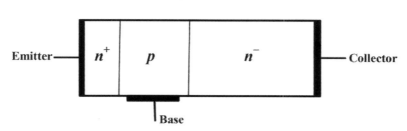

Figure 3.3. Structure of *npn*-BJT.

Figure 3.4 shows the band diagram of an *npn*-BJT at equilibrium; there are two *pn*-junctions. An interesting question is often raised: is a BJT equivalent to two *pn*-junction diodes put together? The short answer is 'no'. When electrons diffuse from the emitter to the collector, they need to pass the *p*-type base region, so it is possible for them to recombine with the large number of holes there. Therefore, in order to prevent this from happening, the base region must be fairly thin. However, if it is too thin, breakdown will happen at a rather low voltage. In the early years of the semiconductor industry, this stringent requirement was a serious challenge in the development of the fabrication process.

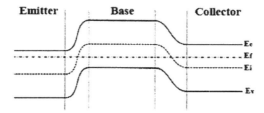

Figure 3.4. Band diagram of *npn*-BJT.

In the characterization process, the emitter is usually grounded, and V_{BE} and V_{CE} can be adjusted. Figure 3.5 shows the circuit diagram and *I–V* characteristics of an *npn*-BJT, which can be divided into three regions. The first region is actually at the horizontal axis, and there is virtually no collector current ($I_C = 0$). From the band

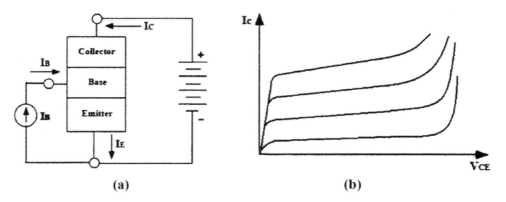

Figure 3.5. (*a*) Common-emitter configuration and (*b*) *I–V* characteristics of *npn*-BJT.

diagram shown in figure 3.4 we can see that the base region looks like a high barrier between the emitter and the collector, which will block the flow of electrons. Therefore, if the barrier height is not brought down by applying a sufficiently high base voltage, the BJT will be in the *cutoff* mode. Please keep in mind that the band diagram shows the energy of negatively charged electrons, so a positive bias voltage will lower the energy level. For a silicon BJT, the threshold voltage of V_{BE} starts from 0.5 V, but 0.7 V is needed to get relatively high current, so this value is used in DC analysis. There are a family of curves in figure 3.5(*b*); each of them corresponds to a specific value of V_{BE}. However, as the diode current is a sensitive exponential function of the bias voltage, the base current is usually specified for each curve.

Horizontally these curves can be divided into two regions. In a very narrow slice on the left-hand side ($V_{CE} < 0.2$ V), the collector currents increase rapidly with V_{CE}; this is called the *saturation* region. When the conduction band in the base region is lowered enough, huge numbers of electrons will diffuse into the base region. However, if the conduction band in the collector region is not sufficiently low, these electrons cannot escape quickly, so they will accumulate in the base region, just as rain water accumulates on top of saturated ground, which is the origin of this terminology. On the other hand, when the position of the conduction band in the collector region is low enough, electrons can escape quickly; however, there is little effect on the collector current when the band of the collector is pulled down further, so the curves become flat, and this is the *active* region. If V_{CE} is increased to very high values, breakdown will happen and this results in the rapid increase of collector current, which is shown on the far right-hand side of these curves.

When a BJT is working in the active mode, the ratio of the collector current over the base current (common emitter current gain β) changes little in a wide range of current levels, which is shown in figure 3.6. On the datasheets of BJTs the DC and AC current gains are often specified as $h_{FE} = \beta_{DC} = \frac{I_C}{I_B}$ and $h_{fe} = \beta_{AC} = \frac{\partial i_C}{\partial i_B} = \frac{i_c}{i_b}$, respectively. Incidentally, these two parameters can be different if the h_{FE} curve is not flat at the Q-point (with a specific I_C). The letter h in these expressions

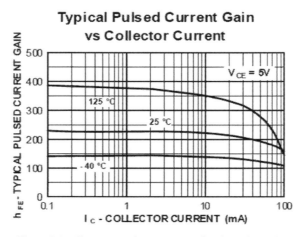

Figure 3.6. Common emitter current gain of 2N3904 BJT.

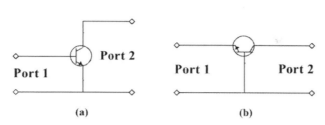

Figure 3.7. Two-port network diagrams of (a) CE and (b) CB configurations.

Table 3.1. Hybrid parameters of BJT.

Meaning	BJT parameter	Two-port network
Input impedance	h_{ie}	$h_{11} = v_1/i_1$
Reverse feedback ratio	h_{re}	$h_{12} = v_1/v_2$
Forward current gain	h_{fe}	$h_{21} = i_2/i_1$
Output admittance	h_{oe}	$h_{22} = i_2/v_2$

stands for *hybrid parameter* in a two-port network, and the subscript e stands for common-emitter configuration in characterization, which is shown in figure 3.7(a). In hybrid configuration, the input signals are the base current (i_1) and collector–emitter voltage (v_2), and the output signals are the base–emitter voltage (v_1) and the collector current (i_2). The four hybrid parameters are listed in table 3.1.

Besides the common-emitter configuration, the triode vacuum tube style common-base configuration, which is shown in figure 3.7(b), was also adopted in characterizing the device in early times. In that case, another set of hybrid parameters can be defined, and two of them are pretty useful: $h_{ib} = \frac{v1}{i1} = \frac{v_{be}}{i_e} = r_e$ (this parameter will be used in the T-model) and $h_{fb} = \frac{i2}{i1} = \frac{i_c}{-i_e} = -\alpha_{AC}$.

Just like most semiconductor devices, the parameters of BJT can vary in a wide range when temperature changes, such as the current gain curves shown in figure 3.6. The drift of these parameters causes a serious problem in electronic circuit design, so negative feedback is used in almost all practical analog electronic circuits.

3.3 Bias circuits

As mentioned in chapter 1, if an amplifier is considered a stringed musical instrument, the bias circuit will be its tuning before playing it. Therefore, the first step in troubleshooting a circuit is checking the bias circuit, and the set of these DC parameters is often called a Q-point. The DC analysis methods are discussed thoroughly in any standard textbook; in addition, the DC parameters of any specific design can be found easily with circuit simulators, so we will emphasize the development of intuition. Nowadays, circuit designers rely heavily on simulation software packages by way of an iterative human–simulator feedback process, as the complexity of most practical circuits is far beyond the mental power of human beings. However, the task of initiation and optimization in the design process falls squarely on the shoulders of engineers, who cannot be replaced by software. Although the fundamental knowledge is very important, intuition is the trait of an expert.

3.3.1 Large signal model of BJT

Based on the specific bias condition, a BJT can be in one of three modes: cutoff, saturation, or active. In most digital circuits—with the exception of ECL (emitter-coupled logic)—the operation of BJTs is switching between cutoff and saturation modes. On the other hand, they need to be in active mode in analog circuits. Therefore, we assume all the BJTs we discuss in this and the following sections are in active mode.

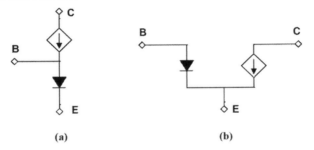

(a)   (b)

Figure 3.8. Large signal model of BJT in (a) common-base and (b) common-emitter configurations.

A simplified large signal model of an *npn*-BJT is shown in figure 3.8, where the base–emitter junction works as a diode, while the collector current is proportional to the emitter or base current and the ratios are the current gains α or β. The Kirchhoff current law (KCL) can be applied to the BJT; the emitter can be considered the source of current, and it branches out into base and collector currents:

$$I_E = I_B + I_C. \tag{3.1}$$

The equation above is valid even in saturation mode, but a larger portion of the emitter current goes to the base. In active mode, the common-emitter current gain β and the common-base current gain α are closely related:

$$\frac{I_C}{\alpha} = \frac{I_C}{\beta} + I_C. \tag{3.2}$$

With the common factor I_C in the equation above canceled, the relationship between α and β can be derived:

$$\alpha = \frac{\beta}{\beta + 1}, \qquad \beta = \frac{\alpha}{1 - \alpha}. \tag{3.3}$$

3.3.2 Primitive bias circuit

Figure 3.9 shows a primitive bias circuit with two independent voltage sources (V_{BB} and V_{CC}). The base–emitter junction works as a silicon diode, so its voltage drop can be assumed to be around 0.7 V. In this way, the base current can be found easily:

$$I_B = \frac{V_{BB} - 0.7}{R_B}. \tag{3.4}$$

With the current gain relationship, I_C can be found next, and then the voltage at the collector node can be derived:

$$V_C = V_{CC} - I_C R_C. \tag{3.5}$$

As mentioned in section 1.4, the ideal value of V_C is in the middle between V_E and V_{CC} so that the output waveforms will not be distorted by touching the floor or ceiling. If one tries to tune this Q-point by adjusting V_{BB}, then one needs to keep in mind that it operates like an inverter circuit: V_C and V_{BB} move in opposite directions. This can be understood from equation (3.5), since there is a negative sign in front of the second term.

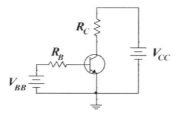

Figure 3.9. Primitive bias circuit.

3.3.3 Bias circuit with emitter degenerate resistor

If β is a constant, then the bias circuit above works fine. However, β changes wildly with temperature. From figure 3.6 we can see that the value of β varies from 140 at $-40\,°C$ to 380 at 125 $°C$ ($I_C = 1$ mA). If the bias circuit is designed with the value of β at a low temperature, V_C will drop when the temperature rises and it is possible even to migrate into saturation mode. In order to overcome such temperature-drift problems, a resistor is added under the emitter with the fancy name 'emitter degenerate resistor'. As the name indicates, it could have some detrimental effects on the circuit.

The analysis of this circuit involves two loops or pathways: the base–emitter loop and the collector–emitter loop. With the KVL, two equations can be set up:

$$I_B R_B + 0.7 + (\beta + 1) I_B R_E = V_{BB} \tag{3.6}$$

$$I_C R_C + V_{CE} + (I_C/\alpha) R_E = V_{CC}. \tag{3.7}$$

The base current can be found from equation (3.6), and then the collector current can be derived next:

$$I_C = \frac{V_{BB} - 0.7}{R_B/\beta + R_E/\alpha}. \tag{3.8}$$

Although β can vary in a wide range, α changes very little. For example, $\alpha = 0.9901$ when $\beta = 100$; $\alpha = 0.9975$ when $\beta = 400$. Therefore, the second term in the denominator of equation (3.8) is very stable; if this is the dominant term, then I_C becomes stable too. Therefore, the requirement for design of a stable bias circuit is $R_E > 5R_B/\beta$. On the other hand, equation (3.7) indicates that R_E can reduce the headroom of swing, $V_{CE} \approx V_{CC} - I_C(R_C + R_E)$, so it is undesirable to choose a very large R_E.

From another point of view, R_E provides negative feedback. Suppose there is a sudden increase of the current gain, $\beta \rightarrow \beta + \Delta\beta$, and it causes an increase of collector current, $I_C \rightarrow I_C + \Delta I_C$, and this in turn causes an increase of the voltage at the emitter, $V_E \rightarrow V_E + \Delta V_E$. From equation (3.6) it can be found that the base current will drop: $I_B \rightarrow I_B - \Delta I_B$. Finally, this makes the collector current less sensitive to the change of β.

3.3.4 Bias circuit with voltage divider

The drawback of the bias circuit shown in figure 3.10 is that it needs two independent voltage sources (V_{BB} and V_{CC}). Unfortunately, usually only V_{CC} is available in practical circuits. Therefore, a voltage divider is used to generate V_{BB} (figure 3.11). There is a common mistake made by beginners in assuming the base voltage is at $V_B = \dfrac{R_{B2}}{R_{B1} + R_{B2}} V_{CC}$; this expression is valid only if I_B is zero or negligible. One way to make this expression valid is using rather low resistors in R_{B1} and R_{B2}, but this will waste a lot of power. Furthermore, in the AC circuit—this will be discussed in the following sections—R_{B1} and R_{B2} contribute to the input resistance, so it is not a good idea to put two small resistors there.

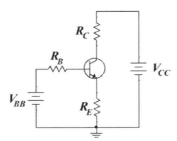

Figure 3.10. Bias circuit with emitter degenerate resistor.

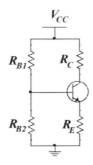

Figure 3.11. Bias circuit with voltage divider.

Actually, with Thévenin's theorem this circuit can be converted into the previous one shown in figure 3.10, and this conversion is straightforward:

$$V_{BB} = \frac{R_{B2}}{R_{B1} + R_{B2}} V_{CC}, \qquad R_B = R_{B1} \| R_{B2} = \frac{R_{B1} R_{B2}}{R_{B1} + R_{B2}}. \qquad (3.9)$$

The design of a bias circuit usually starts with selecting I_C, which determines the transconductance and voltage gain. For a circuit with fairly high immunity to temperature drift, the voltage on the right-hand branch is equally partitioned: $V_E = \frac{1}{3} V_{CC}$ and $V_C = \frac{2}{3} V_{CC}$. If the requirement is not very high, the following partition plan can be adopted: $V_E = \frac{1}{5} V_{CC}$ and $V_C = \frac{3}{5} V_{CC}$. With I_C, V_E, and V_C selected, R_C and R_E can be calculated easily with Ohm's law. With these parameters available, the two resistors on the bias branch can be designed. First, I_B and V_B can be found easily from I_C and V_E, respectively. Second, R_B can be figured out by using a rule of thumb: $R_B = \frac{\beta}{5} R_E$ (moderate stability) or $R_B = \frac{\beta}{10} R_E$ (higher stability), and then $V_{BB} = V_B + I_B R_B$. Finally, the value of R_{B1} and R_{B2} can be calculated with equation (3.9).

3.3.5 Bias circuit with collector–base feedback resistor

Another useful bias circuit involves a resistor between base and collector, which is shown in figure 3.12. R_B falls on the feedback path between the output and input nodes, so it makes the bias circuit more stable, but the drawbacks include a decrease in voltage gain and lower input resistance. In addition, the BJT in this circuit is always in active mode and there is no worry about drifting into saturation mode. Furthermore, the design of this bias circuit is straightforward.

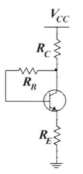

Figure 3.12. Bias circuit with collector–base feedback resistor.

The key equation for this circuit can be derived by following this pathway: $V_{CC} \rightarrow R_C \rightarrow R_B \rightarrow V_{BE} \rightarrow R_E$.

$$I_E R_C + I_B R_B + 0.7 + I_E R_E = V_{CC}. \qquad (3.10)$$

One needs to pay attention to the fact that the current going through R_C is not I_C, but $I_E = I_C + I_B$ instead; this can be derived by applying the KCL at the collector node. If I_E, V_E, and V_C are specified, then R_E, R_C, and R_B can be figured out easily.

3.4 Small signal models

With DC analysis the Q-point of biasing is determined, and the next step is AC analysis. As mentioned in chapter 1, some transformations are needed, such as the capacitors and DC sources. In addition, the transistors in the circuit should also be replaced by their small signal models. In this section we will cover two popular models at low frequency.

3.4.1 T-model

As discussed in the early sections, BJT was developed following the idea of the triode vacuum tube, so initially the common-base (CB) configuration was very popular. The CB large signal model is shown in figure 3.8(a), and its transition to the small signal model is straightforward: replacing the diode with its incremental resistance r_e. Figure 3.13 shows a simplified small signal T-model of BJT.

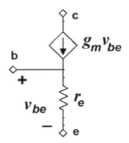

Figure 3.13. Small signal T-model of BJT.

There are two parameters in this T-model; the emitter incremental resistance (r_e) and transconductance (g_m) are defined as

$$r_e = \frac{V_T}{I_E}, \qquad g_m = \frac{I_C}{V_T}. \tag{3.11}$$

The $V_T = kT/e$ in the expressions above is the thermal voltage, and it is equal to 25.9 mV at room temperature. If the input signal is the voltage across the base and the emitter v_{be}, then the emitter current can be found:

$$i_e = \frac{v_{be}}{r_e} = \frac{I_E}{V_T} v_{be}. \tag{3.12}$$

With the relationship of CB current gain α, the collector current can be derived:

$$i_c = \alpha i_e = \frac{\alpha I_E}{V_T} v_{be} = g_m v_{be}. \tag{3.13}$$

For discrete BJTs the I–V curves are fairly flat in the active region, and the T-model is a good approximation. However, for BJTs in modern ICs the base region is rather thin and the slope in the I–V curves is much larger, so a resistor should be added between the collector and the emitter, which makes this model rather awkward. Therefore, the hybrid-π model is more popular in such situations.

3.4.2 Hybrid-π model

The common-emitter (CE) large signal model is shown in figure 3.8(b); by replacing the diode with its incremental resistance, a simplified hybrid-π model can be created (figure 3.14). The expression of incremental resistance can be found by following its definition:

$$r_\pi = \frac{V_T}{I_B}. \tag{3.14}$$

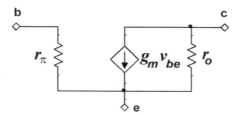

Figure 3.14. Hybrid-π model of BJT.

As I_B is much smaller than I_E, r_π is much higher than r_e:

$$r_\pi = (\beta + 1)r_e. \tag{3.15}$$

If the input is a voltage signal, the base is a better port than the emitter, as the input resistance is higher. On the other hand, for a current signal the emitter is a better option for input port. It is not surprising to see that the current source has the same expression as the T-model in figure 3.13(b), and the derivation procedure is also similar:

$$i_c = \beta i_b = \beta \frac{v_{be}}{r_\pi} = \frac{\beta I_B}{V_T} v_{be} = g_m v_{be}. \tag{3.16}$$

From another point of view, this small signal model can be directly derived from the I–V curves. In the CE configuration, v_{BE} and v_{CE} can be considered two stimulus signals, and the collector current is the response. Therefore, a 3D diagram can be created with the x- and y-axes representing the stimuli v_{CE} and v_{BE}, and with the z-axis showing the response of i_C. The variation (AC signal) of i_C can be determined in this way:

$$\Delta i_C = \frac{\partial i_C}{\partial v_{BE}} \Delta v_{BE} + \frac{\partial i_C}{\partial v_{CE}} \Delta v_{CE}. \tag{3.17}$$

If v_{CE} is kept constant and the perturbation is in v_{BE}, the response is from the exponential function of the diode current amplified by β, so it is very strong:

$$\frac{\partial i_C}{\partial v_{BE}} = \frac{I_C}{V_T} = g_m. \tag{3.18}$$

On the other hand, if v_{BE} is kept constant and the perturbation is in v_{CE}, the response is from the very small slope of the I–V curve, so it is very weak:

$$\frac{\partial i_C}{\partial v_{CE}} = \frac{I_C}{V_A} = \frac{1}{r_o}. \tag{3.19}$$

In this equation V_A is called the Early voltage, named after James M Early, who found that all the I–V curves of a BJT converge to one point on the negative v_{CE} axis. V_A is usually high—it is $\sim$100 V for discrete BJTs—so I_C/V_A can be approximated as the slope of the I–V curve. As we know, a straight line in an I–V diagram can be described by an AC resistance, and its value is equal to the inverse of the slope. Therefore, the output resistance r_o is defined as the inverse of the slope of a specific I–V curve. Thus, the current source and the output resistance on the right-hand side of the hybrid-π model are the response to the AC components of v_{BE} and v_{CE}, and the resulting collector current is

$$i_c = g_m v_{be} + v_{ce}/r_o. \tag{3.20}$$

These two small signal models are derived from the *npn*-BJT, but they can also be used for the *pnp*-BJT without revision. This seems a little counter-intuitive at first, as they look like mirrored structures and the signs are supposed to be flipped. This can be clarified with the perturbation approach. We need to keep in mind that the emitter of a *pnp*-BJT has a higher voltage and the DC current is flowing from the emitter to the base and collector. However, in the small signal model, the emitter is at the bottom and the current is flowing from the base and collector towards the emitter. Figuratively speaking, the small signal model of a *pnp*-BJT is flipped vertically. When a positive perturbation Δv_B is applied, v_{EB} becomes smaller, and then the collector current should decrease, which is equivalent to a superposed current component $g_m \Delta v_B$ flowing *backwards* from the collector to the emitter, and this AC component becomes positive in the small signal model. In the same way, the response to Δv_C is another backward collector current component $\Delta v_C/r_o$.

When the operating frequency increases, the internal capacitors of the BJT need to be taken into account, and they should be included in the small signal models. We will discuss this in detail in the next chapter, when the MOSFET is covered.

3.5 Basic amplifier circuits

For amplifiers with a single discrete BJT, usually one of its three terminals is AC grounded. The ground symbol usually appears at several locations in a circuit, but it is a shared *common* node. Therefore, these three basic configurations are called common-emitter (CE), common-base (CB), and common-collector (CC) amplifiers.

3.5.1 CE amplifier

The CE amplifier is the most useful configuration, since there are gains in both voltage and current. Figure 3.15(*a*) shows an example of a CE amplifier, and all the coupling capacitors are assumed to be large. In DC analysis, these capacitors block the flow of DC current, so only the central part of the circuit remains, which was analyzed in section 3.3. In AC analysis, the large coupling capacitors become short circuits, the node at V_{CC} is grounded, and the BJT is replaced with its small signal model. Then the resulting small signal circuit is shown in figure 3.15(b).

As shown in figure 3.15(a), this circuit can be divided into three sections: signal source (v_{sig} and R_{sig}), core amplifier, and load (R_L). In this way it can be analyzed with the approach of the generic amplifier discussed in section 1.5. First, we need to find the three key parameters of the amplifier: A_{vo}, R_i, and R_o. The intrinsic voltage gain A_{vo} can be found with the two-step approach discussed in section 1.4: $V \rightarrow I$ and $I \rightarrow V$.

$$i_a = g_m v_i, \qquad v_a = -i_a \cdot (r_o \| R_C), \qquad A_{VO} = \frac{v_a}{v_i} = -g_m \cdot (r_o \| R_C). \qquad (3.21)$$

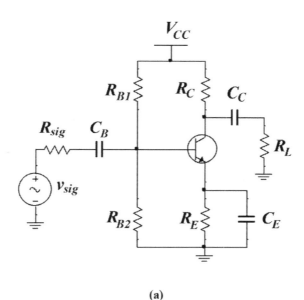

(a)

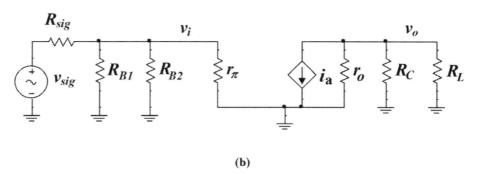

(b)

Figure 3.15. (a) CE BJT amplifier and (b) its small signal circuit.

The input and output resistances can be figured out directly from the small signal circuit:

$$R_i = R_{B1} \| R_{B2} \| r_\pi, \qquad R_o = r_o \| R_C. \qquad (3.22)$$

With equation (1.8)—the formula of the generic amplifier—the gain of this amplifier can be found:

$$A_V = \frac{v_o}{v_{sig}} = -\frac{R_i}{R_{sig} + R_i} g_m R_o \frac{R_L}{R_o + R_L}. \qquad (3.23)$$

3.5.2 CB amplifier

The CB configuration is widely used in multistage amplifiers and in ICs, but it is not a good choice for a single stage amplifier, since its input resistance is very low. Figure 3.16(a) shows an example of a CB amplifier, and its DC section is identical to that of the CE amplifier in figure 3.15(a). Once again, we can assume all the capacitors are large, so they disappear from the small signal circuit, which is shown in figure 3.16(b). In addition, r_o is ignored in the T-model.

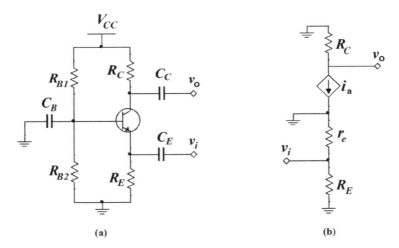

(a)   (b)

Figure 3.16. (a) CB BJT amplifier and (b) its small signal circuit.

This circuit can be analyzed with the same procedure as the CE amplifier. The voltage gain of the core amplifier can be derived in the same way:

$$i_a = g_m v_{be} = -g_m v_i, \qquad v_a = -i_a R_C, \qquad A_{VO} = \frac{v_a}{v_i} = g_m R_C. \qquad (3.24)$$

The input and output resistances can be figured out easily:

$$R_i = R_E \| r_e \approx r_e, \qquad R_o = R_C. \qquad (3.25)$$

Finally, the gain of the complete amplifier circuit can be found in the same way:

$$A_V = \frac{v_o}{v_{sig}} = \frac{R_i}{R_{sig} + R_i} g_m R_o \frac{R_L}{R_o + R_L}. \qquad (3.26)$$

The voltage gain expression for a CB amplifier is very similar to that of a CE amplifier; the only difference is the missing negative sign. However, given the same parameters, the voltage gain of the CB amplifier is much lower, as the input resistance is very low. In addition, there is no current gain in the CB amplifier. Therefore, if the input signal is very weak—such as those from sensors—the CB configuration is not a good option.

3.5.3 CC amplifier

The CC amplifier has a figurative nickname: emitter follower. It indicates that the output voltage at the emitter follows the input voltage at the base. This can be understood easily, as the base–emitter voltage is basically maintained as a constant, so the AC signals at the base and the emitter should be approximately the same. Figure 3.17 shows a CC amplifier circuit; if R_{sig} in the circuit causes some trouble in understanding the follower, it can be ignored for the moment. By fixing the base–emitter junction voltage at $v_{BE} = 0.7\,\text{V}$, the output node should go up and down following the input signal.

Detailed analysis shows that the voltage gain is a little less than unity. This should not be surprising; for example, at the peak of the input signal, the output signal $v_o(t)$ is raised to a higher value, so the current going through R_L rises. Where does this additional current come from? The only source is the emitter of the BJT, so there is an AC component of the emitter current that oscillates with the input signal. This is possible only if the base–emitter junction voltage varies accordingly, but the amplitude is very small due to the sharp exponential I–V relationship of the pn-junction. Therefore, the input signal is divided into two parts: a tiny part falls on the base–emitter junction and most of it falls on the load resistor R_L, so the voltage gain is very close to unity.

Now let us analyze the circuit with the small signal model, which is shown in figure 3.17(b). In order to simplify the procedure, r_o in the T-model is ignored again.

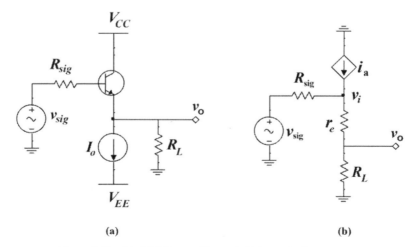

(a) (b)

Figure 3.17. (a) CC BJT amplifier and (b) its small signal circuit.

One approach in analyzing this circuit involves the so-called *resistance reflection rule*, which is used to deal with situations with a transition of current. In this circuit the current in the input branch is i_b, while the current in the output branch is i_e, and they are related by $i_e = (\beta + 1)i_b$. The contribution of the current source at the collector node makes the transition of current possible, but it should be eliminated after the resistances are transformed. Figure 3.18 shows the equivalent circuits looking from the base and emitter ports, respectively.

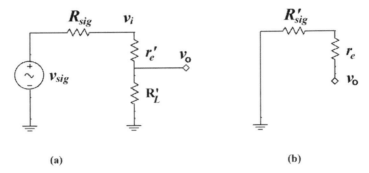

Figure 3.18. Resistance reflection rule: (*a*) base port and (*b*) emitter port equivalent circuits.

In the transformed circuit using the base current in figure 3.18(*a*), r_e and R_L are transformed into r_e' and R_L'. The principle of this transformation is that the voltage should remain the same:

$$i_b r_e' = i_e r_e, \qquad i_b R_L' = i_e R_L. \qquad (3.27)$$

Therefore, the relationship between the resistors is inversely proportional to that of the currents:

$$r_e' = (\beta + 1)r_e, \qquad R_L' = (\beta + 1)R_L. \qquad (3.28)$$

Now the voltage gain can be found with the formula of a voltage divider, and it can be understood that the input signal is divided into three parts with the bulk part falling on the load resistor:

$$A_V = \frac{v_o}{v_{sig}} = \frac{R_L'}{R_{sig} + r_e' + R_L'} = \frac{(\beta + 1)R_L}{R_{sig} + (\beta + 1)(r_e + R_L)}. \qquad (3.29)$$

Usually $R_L \gg r_e$ and $(\beta + 1)R_L \gg R_{sig}$, so the voltage gain is close to unity. Similarly, the input resistance can be found:

$$R_i = r_e' + R_L' = (\beta + 1)(r_e + R_L). \qquad (3.30)$$

In a similar way, the output resistance can be figured out with the transformed circuit shown in figure 3.18(*b*). With the same rule, the source resistance is transformed:

$$i_b R_{sig} = i_e R_{sig}' \quad \rightarrow \quad R_{sig}' = R_{sig}/(\beta + 1). \qquad (3.31)$$

Then the output resistance is simply the sum of the two resistors:

$$R_o = r_e + R_{sig}/(\beta + 1). \qquad (3.32)$$

As discussed in section 1.5, good amplifiers have three requirements: high voltage gain, high input resistance, and low output resistance. Although a CC amplifier does not have any voltage gain at all, the other two conditions are satisfied. Therefore, it can be used as a buffer between two amplifier stages, or between an amplifier and it source/load. From another point of view, it is a current amplifier. In addition, it can also be considered the simplest power amplifier with good linearity and wide bandwidth.

3.6 Amplifiers with feedback

All semiconductor devices are very sensitive to temperature change, and amplifiers with transistors usually need feedback circuits to reduce such sensitivity. Generally speaking, feedback refers to the existence of a pathway for the output signal to return and superpose onto the input signal. There are many other benefits of introducing negative feedback into amplifier circuits, including improvement in linearity and lower noise level, as well as extended bandwidth. However, some unintended feedback paths are undesirable, such as the intrinsic capacitive coupling between the collector and the base.

3.6.1 Amplifier with collector–base feedback resistor

Figure 3.19(a) shows a simple BJT amplifier with a collector–base feedback resistor, and the DC analysis of a similar circuit was covered in section 3.3. In order to concentrate on the feedback effects and simplify the analysis, the source and load are removed. Figure 3.19(b) shows the small signal circuit, and r_o is ignored too for simplification. The resistor R_B looks like a bridge between the input and output, so it provides the feedback path.

This circuit is fairly simple and it can be analyzed with the KCL at the collector node, and the voltage gain can be found easily:

$$\frac{v_o - v_i}{R_B} + g_m v_i + \frac{v_o}{R_C} = 0 \qquad (3.33)$$

$$A_{vo} = \frac{v_o}{v_i} = -\left(g_m R_B - 1\right)\frac{R_C}{R_B + R_C} \approx -g_m(R_B \| R_C). \qquad (3.34)$$

If R_B is much larger than R_C, which is often the case, then the feedback effect is rather weak. In order to develop some intuition and generalize the feedback resistor to any impedance, we can analyze this circuit in a different way. Imagine R_B is a thin rod of uniform resistive conductor with a length of 1 m, and then the voltage at any point on the rod can be measured experimentally. Suppose the internal voltage gain is $A_{VO} = v_o/v_i = -99$; 101 electrodes are equally spaced along the rod so that the voltage distribution can be measured. When a voltage signal v_i is applied at the input

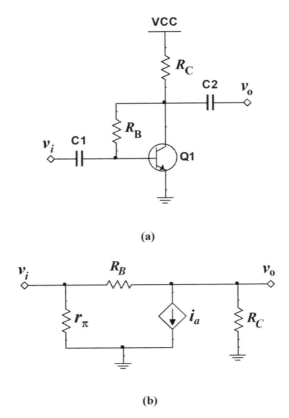

(a)

(b)

Figure 3.19. Amplifier circuits with collector–base feedback resistor.

node, there is a linear distribution of voltage along the rod. Interestingly, the voltage at the second electrode on the left is always zero regardless of the input voltage. Figuratively speaking, the signal distribution along the rod looks like a seesaw, and the position of the second electrode is at the fulcrum point. Therefore, R_B can be partitioned into two resistors at this point, R_{B1} and R_{B2}, with the node between them grounded. The resulting circuit is shown in figure 3.20(a).

As the node between R_{B1} and R_{B2} is grounded, they can be separated, and the transformed circuit is shown in figure 3.20(b). Now the feedback path disappears, so it can be analyzed more easily.

$$A_{VO} = -g_m (R_{B2} \| R_C). \tag{3.35}$$

In general, the voltage gain is assumed to be $-K$, and then the rod can be partitioned in this way: on the left-hand side of the fulcrum point the length is one unit, and on the right-hand side the length is K units. Therefore, the values of R_{B1} and R_{B2} can be found easily:

$$R_{B1} = \frac{R_B}{K+1}, \qquad R_{B2} = \frac{K}{K+1} R_B. \tag{3.36}$$

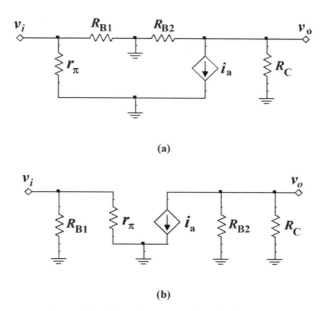

(a)

(b)

Figure 3.20. Transformation of feedback resistor.

If the voltage gain is pretty high, then $R_{B2} \approx R_B$, so there is a good agreement between equation (3.34) and equation (3.35). Usually R_B is much larger than R_C, so the reduction of the gain is insignificant. However, since R_{B1} is much lower than R_B, it will lower the input resistance, which is the major drawback of this circuit.

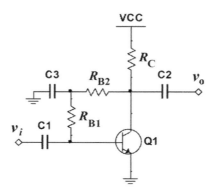

Figure 3.21. Revised amplifier circuit with collector–base feedback resistor.

Fortunately, there is a remedy to the issue of reduced input resistance. Figure 3.21 shows a revised circuit, where R_B is partitioned into R_{B1} and R_{B2} by the designer. For example, an equal partition scheme can be adopted: $R_{B1} = R_{B2} = R_B/2$. As $R_B \gg R_C$ and $R_B \gg r_\pi$, the impact of both R_{B1} and R_{B2} is negligible. In addition, the DC circuit is identical regardless of any partition scheme.

The transformation formula in equation (3.36) can also be extended to situations when other circuit elements serve as the bridge, and the formula of such a general transformation is called *Miller's theorem*:

$$Z_1 = \frac{Z}{K+1}, \qquad Z_2 = \frac{K}{K+1}Z. \qquad (3.37)$$

Inside the structure of BJTs there is an intrinsic coupling capacitor (C_μ) between base and collector, and the following transformation formula is very useful:

$$C_1 = (K+1)C_\mu, \qquad C_2 = \frac{K+1}{K}C_\mu. \qquad (3.38)$$

This expression shows the harmful effect of C_μ, because it is amplified into a very large capacitor (C_1) and will cause a pole at low frequency. However, CB and CC amplifiers do not suffer from this problem, since one side of C_μ is grounded. Therefore, these two kinds of amplifier can work at higher frequencies than CE amplifiers. More details on frequency response will be discussed in the next chapter.

3.6.2 Amplifier with emitter degenerate resistor

The circuit with collector–base feedback resistor can lower the input resistance, but the circuit shown in figure 3.22(a) can increase the input resistance. There is a clue to this difference: at the input node the feedback resistor R_B in figure 3.19(a) is in parallel, while the resistor R_{E1} in figure 3.22(a) is in series with the BJT. In the same way, the feedback can also change the output resistance in two different ways, depending on how the feedback path is connected to the output port.

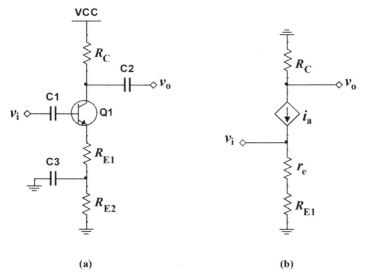

Figure 3.22. (a) Amplifier with emitter degenerate resistor and (b) its small signal circuit.

If this circuit is compared with the circuit shown in figure 3.15, it can be found that the emitter degenerate resistor is split into two: R_{E1} and R_{E2}. As long as the sum of these two resistors is the same as the original one, the DC analysis will be identical. In addition, the position of the capacitor C3 is moved down to the node between R_{E1} and R_{E2}, so R_{E1} shows up in the small signal circuit in figure 3.22(b). In order to simplify the analysis, the source and load are stripped off. In addition, r_o is also ignored in the small signal model. In this way, the voltage gain can be found easily:

$$i_a = \alpha i_e = \frac{\alpha v_i}{r_e + R_{E1}}, \qquad v_o = -i_a R_C,$$

$$A_{VO} = \frac{v_o}{v_i} = -\frac{\alpha R_C}{r_e + R_{E1}} = -\frac{g_m R_C}{\alpha + g_m R_{E1}} \approx -\frac{g_m R_C}{1 + g_m R_{E1}}. \tag{3.39}$$

Next, we can find the input resistance with the *resistance reflection rule*:

$$R_i = (\beta + 1)(r_e + R_{E1}) = r_\pi + (\beta + 1)R_{E1} \approx r_\pi\left(1 + g_m R_{E1}\right). \tag{3.40}$$

Compared with the circuit without partitioning R_E, the voltage gain of this circuit is lowered by a factor of $1 + g_m R_{E1}$, and the input resistance is increased by the same factor, which is called the *amount of feedback*.

The small signal circuit shown in figure 3.22(b) can be understood in a figurative way. A resistor (r_e or R_{E1}) is analogous to a spring, and the conductance—inverse of resistance—is then equivalent to the spring constant or stiffness; the applied voltage signal v_i is equivalent to the displacement, and the current i_e is equivalent to the force in response. If two springs are put in series, the applied displacement is divided between these two springs, so the responsive force will be weaker. In the same way, the applied voltage v_i is now partitioned by r_e and R_{E1}, but the collector current is determined by the voltage across r_e only: $i_a = g_m v_{be} \approx g_m v_i/\left(1 + g_m R_{E1}\right)$. From another point of view, the effective transconductance is reduced by a factor of $1 + g_m R_{E1}$.

Chapter 4

MOSFET amplifier circuits

The concept of the FET—field effect transistor—was patented by Julius Lilienfeld in 1925. Three decades later, the MOSFET was finally invented in 1959 by Dawon Kahng and Martin Atalla at Bell Labs. An early name of the MOSFET was IGFET (insulating gate FET), which highlighted the major difference from the BJT: the gate is insulated from the conducting channel. Although the BJT was invented a decade earlier, MOSFET ICs started to show advantages in high density and low cost in the late 1970s; now they are the workhorse in the microelectronic industry. Historically depletion mode MOSFETs were widely used, but they have been replaced by enhancement mode MOSFETs. Therefore, only this type of MOSFET will be discussed here.

4.1 Introduction to MOSFETs

Just like the BJT, there are two different kinds of MOSFET. The n-channel MOSFET (n-MOS) is similar to an npn-BJT, while the p-channel MOSFET (p-MOS) is similar to a pnp-BJT. Figure 4.1 shows the structure of an n-channel MOSFET. At the center there are three layers of different materials: a metal gate at the top, a silicon dioxide insulator layer below it, and the bulk silicon substrate at the bottom. *MOS* is the abbreviation of the names of these materials. If looked at horizontally, the structure of the three regions with different dopings ($n^+ - p - n^+$) is very similar to that of an npn-BJT. However, the MOSFET is a symmetric device, so the *source* node and the *drain* node can be exchanged.

At the very bottom of the device structure is the body contact, which controls the electric potential of the silicon substrate. Therefore, the traditional MOSFET is actually a four-terminal device. Besides the influences from the gate and drain nodes, there is a *body effect* on the current. Actually, it should be called the *source effect*, as the body of an n-MOS is usually grounded and the body of a p-MOS is directly

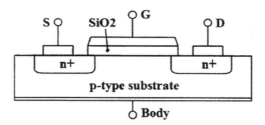

Figure 4.1. Structure of an *n*-channel MOSFET.

connected to V_{DD}, but the source node can be at a different voltage. However, as the source voltage is used as the reference for gate and drain voltages, the relative body voltage seems to be able to change. In such a situation the so-called *body effect* will cause a shift of threshold voltage. Nowadays SOI (silicon-on-insulator) technology has become the mainstream in ICs, as in the UTB or FinFET device structures, and the bulk silicon substrate in the traditional MOSFETs has become an isolated island and is no longer directly connected to an external node. Therefore, these MOSFETs operate in a similar way to the three-terminal BJTs.

We can recall the band diagram of an *npn*-BJT, shown in figure 3.4, where the *p*-type silicon region forms a barrier to the flow of electrons. In MOSFETs the situation is similar, and current cannot flow without appropriate bias voltage on the gate. When a positive voltage is applied to the gate of an *n*-MOS, the energy band of a thin layer of silicon below the gate oxide will be pulled down and a conducting channel can form, and then the flow of electrons between source and drain will be enhanced, so it is called an *enhancement mode* MOSFET.

One way to understand the operation of MOSFETs is using the superposition principle: the potential energy of an electron in the conducting channel is affected by the voltages at all of the four nodes. Let us make things simple by assuming both the *body* and *source* nodes of an *n*-MOS are grounded, and the voltages at the gate and drain nodes can vary. As the gate layer and the conducting channel are only separated by a thin layer of insulator, the influence of the gate voltage is very strong. However, the influence from the drain node is much weaker. There are two reasons: first, the channel length is usually much larger than the thickness of the gate oxide layer; second, silicon is a semiconductor and the freely moving carriers can screen the voltage change at the drain node. Therefore, when the drain voltage is beyond a certain value, it has little effect on the current. The $I-V$ characteristics of an *n*-channel MOSFET are shown below (figure 4.2); the curves correspond to different gate voltages.

Similarly to the $I-V$ characteristics of an *npn*-BJT, they can also be divided into three regions: cutoff, triode, and active. When the gate voltage is not sufficiently high ($V_{GS} < V_{tn}$), the current can be assumed to be zero, so it is called *cutoff* mode, which is the same terminology as used with a BJT. If the gate voltage is high enough, but the drain voltage is rather low, the current changes rapidly with the drain voltage. The shape of the curves in this region is similar to those of a triode vacuum tube shown in figure 3.2, so the operation in this region is said to be in *triode* mode. For a

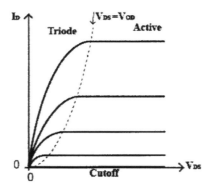

Figure 4.2. $I–V$ characteristics of an n-channel MOSFET.

conventional n-MOS—the channel length is not very short—the current can be described by the following equation:

$$i_D = k_n \left[(v_{GS} - V_{tn}) - \frac{1}{2} v_{DS} \right] v_{DS} = k_n \left(v_{OD} v_{DS} - \frac{1}{2} v_{DS}^2 \right). \tag{4.1}$$

Here V_{tn} stands for the threshold voltage, and the n-MOS would be in cutoff mode if the gate voltage were less than this. In addition, $v_{OD} = v_{GS} - V_{tn}$ is called the *overdrive voltage*, and describes how far the gate voltage is above the threshold voltage.

When the drain voltage is very low, the $I–V$ curves are linear, and basically the conducting channel behaves like a resistor with its resistance controlled by the gate voltage. If the conducting channel is considered as a canal, this overdrive voltage determines the depth of the water there. In this situation, the second term in equation (4.1) can be neglected, and then this equation looks like Ohm's law and $k_n v_{OD}$ becomes the conductance:

$$i_D = (k_n v_{OD}) v_{DS}. \tag{4.2}$$

When the drain voltage is increased beyond the overdrive voltage ($v_{DS} > v_{OD}$), its influence on the conducting channel diminishes and the $I–V$ curves become saturated or level off, so it is said to be in the saturation mode. However, in order to avoid confusion with the saturation mode of the BJT, *active mode* is the preferred name for this region, and the current can be found in a simple expression:

$$i_D = \frac{1}{2} k_n v_{OD}^2. \tag{4.3}$$

The boundary between the triode mode and the active mode is at $v_{DS} = v_{OD}$, so the two equations (4.1) and (4.3) should give the same result at this point, which can be verified easily.

Just like BJTs, in the active mode the drain–source voltage v_{DS} still has some weak influence on the current. When v_{DS} increases, the space charge region of the pn-junction around the drain region becomes wider, then the effective channel length will decrease, which causes the current to increase slightly. This effect can be described with the following equation:

$$i_D = \frac{1}{2}k_n v_{OD}^2 (1 + \lambda v_{DS}).$$ (4.4)

The parameter λ is equivalent to the inverse of the Early voltage for a BJT: $\lambda = 1/V_A$.

For IC designers the parameter k_n can be designed by selecting different device dimensions:

$$k_n = \mu_n C_{ox} \frac{W}{L}.$$ (4.5)

In this expression μ_n stands for the electron mobility in the conducting channel, which is usually lower than the value for bulk silicon with the same level of doping, since the defects at the interface cause more scattering for carriers. The parameter C_{ox} is the specific capacitance of the gate oxide, which is determined by the IC technology with the specific thickness and relative permittivity of the oxide layer. These two parameters are fixed, but the designers can change the width (W) and length (L) of the MOSFETs. In this way, different k_n can be selected for the MOSFETs in a circuit. For example, if higher current is needed, W can be designed to be wider. Usually the shortest L is used in digital circuits to achieve high speed, but longer L is often selected in analog circuits to suppress the *drain-induced barrier lowering* (DIBL) effect.

Besides being the dominant device in ICs, MOSFETs are also widely used in power electronics. In order to accommodate the flow of large currents, the structure of a power MOSFET is different. There are a couple of few alternative configurations: V-MOS, D-MOS, HEXFET, etc. In these devices the source and drain are not interchangeable, so they need to be identified clearly when inserted into a circuit. Although MOSFETs have the advantage of an insulated gate over BJTs, they also have some drawbacks. For example, in MOSFETs the current flows laterally in a thin sheet of conducting channel; on the other hand, in BJTs the current flows *vertically* through the base layer: in this way they can handle large current more efficiently. There is a hybrid device called an IGBT (insulated-gate bipolar transistor), which is a combination of the MOSFET and the BJT. It works better than either type of device individually in power electronics.

4.2 Small signal models

Low frequency small signal models of MOSFETs are derived from the $I-V$ characteristics. Because MOSFETs and BJTs have similar $I-V$ characteristics, their small signal models should also be similar. However, there is a major difference: no gate current in MOSFETs. Therefore, the r_π in the hybrid-π model for BJTs becomes open circuit for MOSFETs. Looking towards the future, the three-terminal MOSFETs will become the mainstream, so the body related effects are ignored.

4.2.1 Hybrid-π model at low frequency

Figure 4.3 shows the low frequency hybrid-π model of MOSFETs. The parameters can be derived in a similar way as for BJTs. As discussed in the previous section, the current is controlled by the gate voltage and drain voltage, where the source voltage is used as the reference. Therefore, an equation similar to equation (3.17) can be formulated:

$$\Delta i_D = \frac{\partial i_D}{\partial v_{GS}} \Delta v_{GS} + \frac{\partial i_D}{\partial v_{DS}} \Delta v_{DS}. \tag{4.6}$$

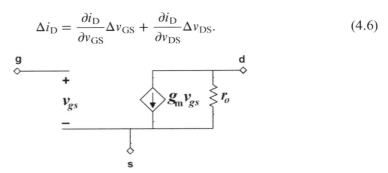

Figure 4.3. Low frequency hybrid-π model of MOSFET.

The rate of current change versus gate–source voltage v_{GS} is defined as the transconductance:

$$g_m = \frac{\partial i_D}{\partial v_{GS}}. \tag{4.7}$$

Similarly, the current change versus drain–source voltage v_{DS} is defined as the output conductance, which is the inverse of the output resistance of this small signal model:

$$r_o = \left(\frac{\partial i_D}{\partial v_{DS}} \right)^{-1}. \tag{4.8}$$

Now the AC drain current can be expressed in this way:

$$i_d = g_m v_{gs} + v_{ds}/r_o. \tag{4.9}$$

The transconductance can be derived from equation (4.3):

$$g_m = \frac{\partial i_D}{\partial v_{OD}} \frac{\partial v_{OD}}{\partial v_{GS}} \bigg|_Q = k_n V_{OD} = \sqrt{2 k_n I_D}. \tag{4.10}$$

The output resistance can be found from equation (4.4):

$$r_o = \left(\frac{\partial i_D}{\partial v_{DS}} \bigg|_Q \right)^{-1} = \frac{1}{\lambda I_D}. \tag{4.11}$$

4.2.2 T-model at low frequency

The T-model of MOSFETs, similar to that of BJTs, is an alternative way to represent the small signal characteristics, which is very convenient to use in some

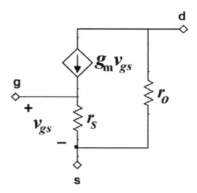

Figure 4.4. T-model of MOSFET.

circuits if the output resistance r_o can be neglected. Figure 4.4 shows the T-mode of a MOSFET with r_o included. At first sight this model does not make sense, as the gate is insulated from the conducting channel. In order to satisfy this condition, the current through the resistance r_s must be equal to that in the current source above it, and from this condition r_s can be derived:

$$\frac{v_{gs}}{r_s} = g_m v_{gs} \rightarrow r_s = \frac{1}{g_m}. \tag{4.12}$$

4.2.3 Hybrid-π model at high frequency

When the operating frequency gets higher, the internal capacitors of MOSFETs need to be taken into account, and a high frequency small signal model is shown in figure 4.5. As we know, there is a capacitive coupling between any two conductors separated in space; in fact, capacitance is the measure of the strength in this coupling. If the coupling to the substrate can be ignored, there are three capacitors between the three nodes of a MOSFET. However, as the source and the drain are separated by conductive materials, the direct coupling between these two nodes (C_{ds}) is much lower than their coupling with the gate (C_{gs} and C_{gd}), so it is ignored.

In addition, the conducting channel underneath the gate is electrically connected to the source, and it is separated from the drain with a *pn*-junction in the active

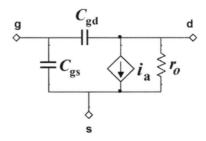

Figure 4.5. High frequency hybrid-π model of MOSFET.

mode. Therefore, the gate–source coupling (C_{gs}) is stronger than the gate–drain coupling (C_{gd}). However, if neither the gate nor the drain is grounded, C_{gd} can be amplified by Miller's effect and the result is a much larger capacitor to the gate node. This will be discussed in detail in section 4.6.

When the frequency gets even higher, other parasitic circuit elements should be included in the small signal model, such as parasitic resistance and inductance, as well as other coupling capacitors. In addition, the parameters of these circuit elements can no longer be expressed in a simple way, and the parameter space has to be partitioned into small domains. Model building is a dynamic process with the relentless progress following Moore's law, and it is very complicated and challenging. Besides the industry standard BSIM models developed by a research group from UC Berkeley, there are a couple of other models, such as Philips MOS models and EKV models (EKV is the abbreviation of the last names of the three developers).

4.3 Basic amplifier circuits

Simple MOSFET amplifiers are also classified in terms of which of the three terminals is AC grounded. Therefore, there are three basic configurations: common-source (CS), common-gate (CG), and common-drain (CD) amplifiers.

4.3.1 CS amplifier

Just like the CE BJT amplifier, the CS MOSFET amplifier is the most commonly used configuration in discrete circuits. Figure 4.6(a) shows an example of a CS amplifier, where all the coupling capacitors are assumed to be large. In AC analysis, the large coupling capacitors (C_G and C_D) and the bypass capacitor (C_S) become short circuits, the node at V_{DD} is grounded, and the MOSFET is replaced with its small signal model. The resulting small signal circuit is shown in figure 4.6(b).

Compared with the CE BJT amplifier circuit, the voltage divider (R_1 and R_2) section is easier to analyze. Since there is no gate current, these two resistors can be very large (~MΩ), and the DC voltage at the gate can be calculated simply with the formula of the voltage divider:

$$V_G = \frac{R_2}{R_1 + R_2} V_{DD}. \tag{4.13}$$

In AC analysis this circuit can be divided into three sections: signal source, core amplifier, and load. The input and output resistances can be found easily:

$$R_i = R_1 \| R_2, \qquad R_o = r_o \| R_D. \tag{4.14}$$

One of the advantages of the MOSFET is the insulated gate, so the input resistance is very high. In this case, the input signal at the gate is very close to the signal

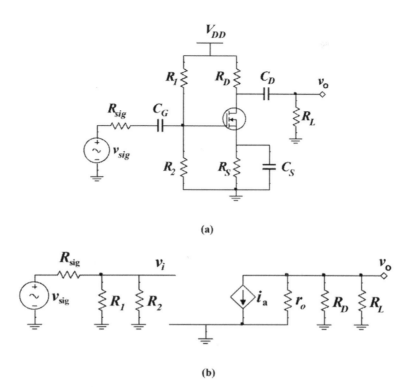

(a)

(b)

Figure 4.6. CS amplifier and its small signal circuit.

from the source: $v_i \approx v_{\text{sig}}$. The remaining part of the analysis is parallel to that of the CE BJT amplifier:

$$i_a = g_m v_i \approx g_m v_{\text{sig}}, \qquad v_o = -i_a \cdot (r_o \| R_D \| R_L),$$

$$A_V = \frac{v_o}{v_{\text{sig}}} = -g_m \cdot (r_o \| R_D \| R_L). \tag{4.15}$$

Although this result is similar to that for the CE BJT amplifier, the gain of the CS MOSFET is usually lower due to the smaller transconductance. A comparison can show this difference:

$$g_m^{\text{BJT}} = \frac{I_C}{V_T}, \qquad g_m^{\text{MOS}} = \frac{2I_D}{V_{\text{OD}}}. \tag{4.16}$$

As we know, $V_T = 25.9mV$ at room temperature, and V_{OD} is usually much higher than V_T at the same current with the discrete MOSFETs commonly used in the laboratory. One question arises: can the MOSFET achieve a higher transconductance if the overdrive voltage V_{OD} is reduced? The trick is that I_D and V_{OD} are not independent; actually $g_m = k_n V_{\text{OD}}$, and thus reducing V_{OD} will cause the transconductance to decrease.

The threshold voltage V_{tn} for the MOSFET is not a hard limit bordering the cutoff mode, and there is a weak current flowing in the subthreshold mode, i.e. $V_{\text{GS}} < V_{\text{tn}}$. Another terminology for this mode is *weak inversion*, which means that the carrier

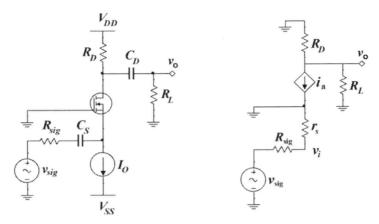

Figure 4.7. CG amplifier and its small signal circuit.

concentration in the conducting channel is rather low. In this case, the gate oxide layer and the weakly inverted p-type layer of silicon underneath are equivalent to two capacitors in series, and then the bending of the energy band in the channel region is strongly affected by the gate voltage, so its behavior is similar to that of a BJT. Working in this mode, the current is very low, but the transconductance—the tangent of the i_D–v_{GS} curve—is pretty high. Therefore, it is the ideal mode for low power designs.

4.3.2 CG amplifier

Figure 4.7 shows a CG amplifier and its small signal circuit. In the analysis of this circuit, the T-model is a better option if r_o is much higher than R_D so that it can be neglected. The input and output resistances are very easy to find:

$$R_i = r_s = 1/g_m, \qquad R_o = R_D. \tag{4.17}$$

The gain of this amplifier can be found in the same procedure as that of a CB BJT amplifier:

$$v_i = \frac{R_i}{R_{sig} + R_i} v_{sig} = \frac{v_{sig}}{1 + g_m R_{sig}}, \qquad i_a = -g_m v_i, \qquad v_o = -i_a (R_o \| R_L) \tag{4.18}$$

$$A_V = \frac{v_o}{v_{sig}} = \frac{g_m (R_D \| R_L)}{1 + g_m R_{sig}}. \tag{4.19}$$

Just like the CB BJT amplifier, there is no current gain in the CG amplifier, and it also suffers from low input resistance. However, if the input signal is from a current source, as in ICs, the low input resistance becomes an advantage, as it can draw more current from the source.

4.3.3 CD amplifier

The CC BJT amplifier has the nickname *emitter follower*; similarly, the CD amplifier is called *source follower*. As the gate is insulated from the conducting channel, the

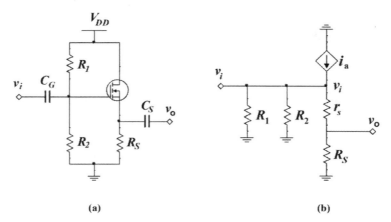

Figure 4.8. CD amplifier and its small signal circuit.

CD amplifier works better than the emitter follower when the source signal is very weak with little output current. Apart from this, the working principle is almost identical. Figure 4.8 shows a CD amplifier; the input and output impedances are

$$R_i = R_1 \| R_2, \qquad R_o = r_s \| R_s = \frac{R_s}{1 + g_m R_s}. \tag{4.20}$$

Compared with a CC BJT amplifier with the same configuration, they are equivalent if $\beta \to \infty$ is assumed.

4.4 First order RC filters

In order to analyze the frequency response of MOSFET amplifier circuits, we need to review first order RC filters. These filters fall into two categories: the low-pass filter (LPF) and high-pass filter (HPF). In amplifier circuits the capacitors can be analyzed individually, but each capacitor is often connected to more than one resistor. These subcircuits are discussed in detail so that the formulae can be used directly in the next two sections.

4.4.1 Phasor expression

In order to understand the $I - V$ relationship of capacitors intuitively, one can imagine a capacitor as a water container. If the capacitance is a constant, then the container looks like a fish tank and the capacitance is equal to the base area. Originally, the terminology *capacitance* was named for the capacity in holding electric charge, just like a tank holding water. The voltage across a capacitor is equivalent to the height of the water level in the tank, and the amount of charge is equal to the volume of the water. Therefore, these three parameters have the following relationship:

$$Q = CV \quad (Volume = Base\ Area \times Height). \tag{4.21}$$

If an inlet pipe is connected to the tank, the flux rate of water in the pipe will be equivalent to the current, and the dynamic relationship can be described by a differential equation:

$$i_C(t) = \frac{dQ}{dt} = C\frac{dv_C(t)}{dt}. \tag{4.22}$$

Even in the case of a variable capacitance, such as a reverse biased *pn*-junction, this differential equation is still valid. A variable capacitor can be visualized as a vase with a variable cross-sectional area at different heights, which is equivalent to the voltage dependent capacitance $C = C(V)$.

Unlike the linear $I-V$ relationship in Ohm's law, such a differential relationship is not very convenient to use in circuit analysis. For example, in most applications the input signal is a harmonic function, but the derivative of a sine function becomes a cosine function and vice versa. Therefore, some transformations are needed. The simple approach is converting a sine function into an exponential function with Euler's formula:

$$e^{j\omega t} = \cos(\omega t) + j\sin(\omega t). \tag{4.23}$$

As the derivative of an exponential function is still the same type of function, a linear relationship can be established. Here the difference between a sine function and a cosine function is reflected by the additional factor j, or a 90° phase shift. For example, $v_C(t) = V_o\cos(\omega t)$ can be converted into an exponential function, $\tilde{v}_C(t) = V_o\exp(j\omega t)$, and then the current can be derived with equation (4.22):

$$\tilde{i}_C(t) = j\omega C\ \tilde{v}_C(t). \tag{4.24}$$

Now there is a linear relationship between current and voltage signals in exponential format, and the admittance and impedance can be defined:

$$Y = j\omega C = sC, \qquad Z = \frac{1}{j\omega C} = \frac{1}{sC}. \tag{4.25}$$

One can replace $j\omega$ by s so that the resulting formula looks more elegant. However, beginners can keep the original formula, which reminds us of the imaginary characteristics.

For linear circuits, the frequency of signals is the same regardless of where they are measured. This is similar to the measurement of one's pulse rate; there is no difference in taking the pulse from the heart or from the arm. Therefore, the component $\exp(j\omega t)$ can be removed and results in the phasor format, which contains the information of magnitude and phase. Equation (4.24) can be transformed into phasor format:

$$\tilde{I}_C = j\omega C\ \tilde{V}_C = sC\tilde{V}_C. \tag{4.26}$$

4.4.2 *RC* LPF

Figure 4.9(*a*) shows an *RC* LPF. Its transfer function—voltage gain in the frequency domain—can be found easily with the formula of a voltage divider circuit, which is shown in figure 4.9(*b*).

$$T(s) = \frac{\tilde{V}_o}{\tilde{V}_i} = \frac{Z_2}{Z_1 + Z_2} = \frac{1/sC}{R + 1/sC} = \frac{1}{1 + sRC}. \tag{4.27}$$

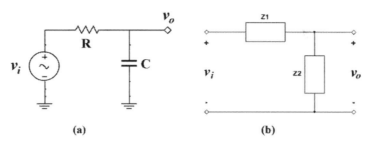

Figure 4.9. *RC* low-pass filter.

The product *RC* has the dimension of time, which is often called the time constant of an *RC* circuit. If the capacitance is imagined as a water tank, then the resistance can be considered as the inlet or outlet pipe with the cross-sectional area inversely proportional to the resistance. Therefore, a large resistor is equivalent to a thin pipe, while a large capacitor corresponds to a large container. Thus, it will take a very long time to fill or drain the tank, and this can be described by the time constant $\tau = RC$. In addition, the inverse of time has the dimension of frequency, so a critical frequency can be defined in this way:

$$\omega_c = \frac{1}{RC}, \qquad f_c = \frac{1}{2\pi RC}. \qquad (4.28)$$

With this expression, equation (4.27) can be rewritten as

$$T(s) = \frac{1}{1 + s/\omega_c} = \frac{\omega_c}{s + \omega_c}. \qquad (4.29)$$

The characteristics of this transfer function can be examined at three domains:

(a) $\omega \ll \omega_c$: $T(s) = 1$;

(b) $\omega = \omega_c$: $T(s) = \dfrac{1}{1 + j} = \dfrac{1}{\sqrt{2}}e^{-j\frac{\pi}{4}} = \dfrac{1}{\sqrt{2}}\angle{-45°}$;

(c) $\omega \gg \omega_c$: $T(s) = \dfrac{\omega_c}{j\omega} = \dfrac{\omega_c}{\omega}e^{-j\frac{\pi}{2}} = \dfrac{\omega_c}{\omega}\angle{-90°}$.

Figure 4.10(*a*) shows this frequency response with a linear scale. When the frequency is much lower than the critical frequency ($\omega < 0.1\omega_c$), the filter seems transparent and signals can pass through without attenuation. This can be understood from the behavior of the capacitor, as it is equivalent to an open circuit at low frequency. At the critical frequency there is an attenuation factor of $1/\sqrt{2} \approx 0.707$ and phase shift of $-45°$. When the frequency is much higher than the critical frequency ($\omega > 10\omega_c$), there is a rapid drop in the magnitude of the transfer function and it is inversely proportional to the signal frequency. On the other hand, the phase shift will saturate at $-90°$.

In most applications, filters are used to reject interference signals, so the *tail* of the transfer function is very interesting. In order to show it more clearly, a logarithm

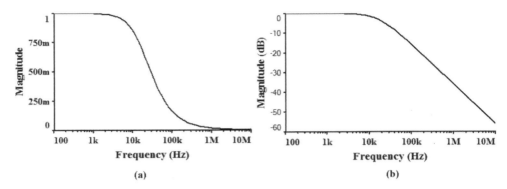

Figure 4.10. Frequency response of *RC* LPF.

scale with the unit of *decibels* is used to represent the magnitude of the transfer function, and a diagram such as shown in figure 4.10(*b*) is called a *Bode plot*—named after Hendrik Alfred Bode, a pioneer of modern control theory and electronic communication.

The unit of *decibels* was originally used in the telephone industry to describe the attenuation of sound, so its definition is related to the ratio of powers:

$$\text{ratio(dB)} = 10 \log_{10}\!\left(\frac{P_2}{P_1}\right). \tag{4.30}$$

With a logarithm function, very large or very small ratios can be expressed in such a way that they are much easier to comprehend. For example, a power ratio of 10^5 can be converted into 50 dB; similarly, a ratio of 10^{-5} can be converted into -50 dB. However, in electronic circuits, the power cannot be measured directly, so the voltage ratio is used instead:

$$\text{ratio(dB)} = 10 \log_{10}\!\left(\frac{V_2}{V_1}\right)^2 = 20 \log_{10}\!\left(\frac{V_2}{V_1}\right). \tag{4.31}$$

At $\omega = \omega_c$, $|T(s)|$ in dB $= -20 \log_{10}(\sqrt{2}) \approx -3$ dB. In a Bode plot there is a sharp bend at this point, so ω_c or f_c is also called the *corner frequency*. For $\omega \gg \omega_c$, $|T(s)|$ in dB $= 20 \log_{10}(\omega_c) - 20 \log_{10}(\omega)$. If the transfer function is calculated at two frequencies with $\omega_2 = 10\,\omega_1$, it can be found that the transfer function drops by 20 dB, so the slope of the transfer function is described as -20 dB/decade. If $\omega_2 = 2\omega_1$, the transfer function will drop by 6 dB, so this slope can also be described as -6 dB/octave.

4.4.3 *RC* HPF

If the locations of the resistor and capacitor in figure 4.9(*a*) are interchanged, the LPF will become an HPF, which is shown in figure 4.11(*a*). Following the same procedure, the transfer function can be derived:

$$T(s) = \frac{\tilde{V}_o}{\tilde{V}_i} = \frac{Z_2}{Z_1 + Z_2} = \frac{R}{R + 1/sC} = \frac{sRC}{1 + sRC} = \frac{s}{s + \omega_c}. \tag{4.32}$$

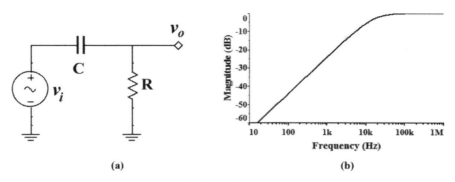

(a)   (b)

Figure 4.11. *RC* HPF and its frequency response.

The characteristics of this transfer function can be examined in three domains:

(a) $\omega \ll \omega_c$: $T(s) = \dfrac{s}{\omega_c} = j\dfrac{\omega}{\omega_c} = \dfrac{\omega}{\omega_c}\angle 90°$;

(b) $\omega = \omega_c$: $T(s) = \dfrac{j}{1+j} = \dfrac{1}{\sqrt{2}}e^{j\frac{\pi}{4}} = \dfrac{1}{\sqrt{2}}\angle 45°$;

(c) $\omega \gg \omega_c$: $T(s) = 1$.

The behavior of the transfer function is shown in figure 4.11(*b*), and is opposite to that of the LPF shown in figure 4.10(*b*). In the same way, the slope of the low frequency region can be described as 20 dB/dec or 6 dB/oct. If one has trouble in distinguishing LPF and HPF, the behavior of the capacitor can be used for identification. In this circuit, the capacitor is equivalent to a short circuit at high frequency, so it must be an HPF.

4.4.4 *RCR* LPF and HPF

In amplifier circuits a capacitor is usually connected to more than one resistor; Figure 4.12 shows an *RCR* LPF and HPF. The method of analysis is the same if they are converted into a voltage divider circuit first. In the circuit shown in figure 4.12(*a*) the admittance of the combination of C and R_2 can be found first:

$$Y_2 = sC + \frac{1}{R_2} = \frac{1 + sCR_2}{R_2}.\tag{4.33}$$

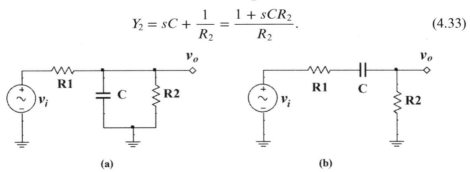

(a)   (b)

Figure 4.12. (*a*) *RCR* LPF; (*b*) *RCR* HPF.

Next the transfer function can be found:

$$T(s) = \frac{\tilde{V}_o}{\tilde{V}_i} = \frac{Z_2}{Z_1 + Z_2} = \frac{1}{1 + Z_1 Y_2} = \frac{R_2}{R_1 + R_2} \frac{\omega_c}{s + \omega_c}. \qquad (4.34)$$

This transfer function has two factors: the first one is a voltage divider transfer function, and the second is an LPF transfer function with the corner frequency at $\omega_c = 1/[(R_1 \| R_2)C]$. At low frequency the capacitor is equivalent to an open circuit, so it is just a voltage divider with R_1 and R_2. Looking out from the capacitor, R_1 and R_2 are in parallel if the input signal source is shorted, which results in the expression for ω_c.

The analysis of the *RCR* HPF circuit in figure 4.12(b) is straightforward: $Z_1 = R_1 + 1/sC$.

$$T(s) = \frac{\tilde{V}_o}{\tilde{V}_i} = \frac{Z_2}{Z_1 + Z_2} = \frac{R_2}{R_1 + R_2} \frac{s}{s + \omega_c}. \qquad (4.35)$$

This transfer function also has two factors: the first one is the same voltage divider transfer function, but the second is an HPF transfer function with the corner frequency at $\omega_c = 1/[(R_1 + R_2)C]$. This can be understood with the behavior of the capacitor, too. At high frequency, the capacitor is equivalent to a short circuit, and then this circuit becomes a voltage divider with R_1 and R_2. Looking out from the capacitor, R_1 and R_2 are in series, resulting in the expression of ω_c. This result can be derived in a different way: first interchange the positions of R_1 and C, let the voltage at the node between them be v_m, and then these two factors can show up clearly:

$$T(s) = \frac{\tilde{V}_o}{\tilde{V}_i} = \frac{\tilde{V}_o}{\tilde{V}_m} \frac{\tilde{V}_m}{\tilde{V}_i} = \frac{R_2}{R_1 + R_2} \frac{s}{s + \omega_c}. \qquad (4.36)$$

4.4.5 *RCR* step-down filter

The circuit shown in figure 4.13(a) looks very similar to the circuit in figure 4.12(b); however, the frequency response is quite different. The transfer function can be derived in the same way:

$$T(s) = \frac{\tilde{V}_o}{\tilde{V}_i} = \frac{Z_2}{Z_1 + Z_2} = \frac{R_2 + 1/sC}{R_1 + R_2 + 1/sC} = \frac{R_2}{R_1 + R_2} \frac{s + \omega_z}{s + \omega_p}. \qquad (4.37)$$

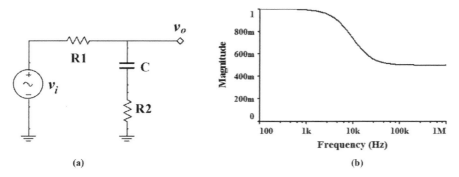

(a) (b)

Figure 4.13. *RCR* step-down filter and its transfer function.

In this equation, $\omega_z = 1/R_2 C$ and $\omega_p = 1/(R_1 + R_2)C$, so $\omega_z > \omega_p$. The characteristics can be analyzed in this way:

(a) $\omega \ll \omega_p$: $T(s) = \dfrac{R_2}{R_1 + R_2} \dfrac{\omega_z}{\omega_p} = 1$;

(b) $\omega \gg \omega_z$: $T(s) = \dfrac{R_2}{R_1 + R_2}$.

The characteristics of this filter can be understood from the behavior of the capacitor. At low frequency it is equivalent to an open circuit, so $v_o(t) = v_i(t)$. At high frequency it is equivalent to a short circuit, so $v_o(t) = \dfrac{R_2}{R_1 + R_2} v_i(t)$. Figure 4.13(b) shows the simulated result of a step-down filter with the following parameters: $R_1 = R_2 = 1\,\mathrm{k}\Omega$, $C = 10\,\mathrm{nF}$. With a simple calculation one can find $f_z = 15.9\,\mathrm{kHz}$ and $f_p = 7.96\,\mathrm{kHz}$.

4.4.6 *RCR* step-up filter

A step-up filter circuit is shown in figure 4.14(a), and it can be analyzed in the same way:

$$T(s) = \frac{\tilde{V}_o}{\tilde{V}_i} = \frac{Z_2}{Z_1 + Z_2} = \frac{Y_1 Z_2}{1 + Y_1 Z_2} = \frac{R_2 + s R_1 R_2 C}{(R_1 + R_2) + s R_1 R_2 C} = \frac{s + \omega_z}{s + \omega_p}. \quad (4.38)$$

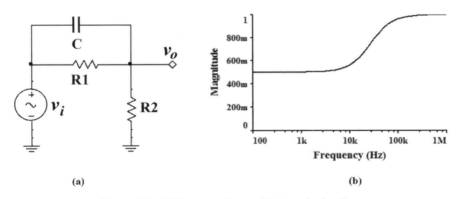

(a) (b)

Figure 4.14. *RCR* step-up filter and its transfer function.

In this equation $\omega_z = 1/R_1 C$ and $\omega_p = 1/(R_1 \| R_2)C$, so $\omega_z < \omega_p$. The characteristics of this filter are opposite to those of the previous one.

(a) $\omega \ll \omega_z$: $T(s) = \dfrac{\omega_z}{\omega_p} = \dfrac{R_2}{R_1 + R_2}$;

(b) $\omega \gg \omega_p$: $T(s) = 1$.

The characteristics of this filter can also be understood from the behavior of the capacitor. At high frequency it is equivalent to a short circuit, so $v_o(t) = v_i(t)$. At low frequency it is equivalent to an open circuit, so $v_o(t) = \dfrac{R_2}{R_1 + R_2} v_i(t)$. Figure 4.13(b) shows the simulated result of a step-up filter with the following parameters: $R_1 = R_2 = 1\,k\Omega$, $C = 10\,nF$. With a simple calculation one can find $f_z = 15.9\,kHz$ and $f_p = 31.8\,kHz$.

4.5 Low frequency response of CS amplifier

As discussed in section 1.5, the generic frequency response of amplifier circuits is shown in figure 4.15. In the low frequency domain the external capacitors form HPFs, and in the high frequency region the internal capacitors form LPFs. If we analyze the circuits in the *midband*, all these capacitors can be ignored, which is what we have done so far. In this section we will concentrate on the low frequency response of the CS amplifier, and the high frequency response will be covered in the next section.

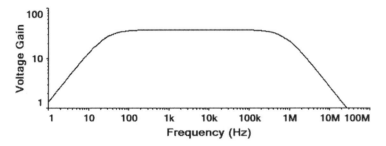

Figure 4.15. Frequency response of amplifier circuits.

The amplifier circuit shown in figure 4.6(*a*) was analyzed in the midband in section 4.3. Now we can keep all the external capacitors and analyze its low frequency response; the small signal circuit is shown in figure 4.16. There are two coupling capacitors (C_g and C_d) and each of them has the transfer function of an HPF. In addition, there is also a bypass capacitor (C_s) and it has the step-up transfer function. If the corresponding corner frequencies are not very close to each other, they can be analyzed separately. Therefore, when the frequency response of one capacitor is analyzed, the others can be assumed to be very large and become short circuits in the small signal circuit.

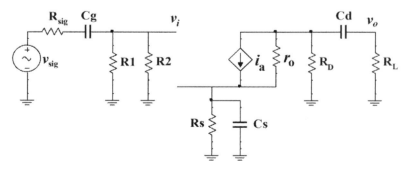

Figure 4.16. Small signal circuit of CS amplifier.

4.5.1 Gate capacitance

Figure 4.17 shows the small signal circuit with C_d and C_s shorted, so the frequency response is determined by the subcircuit with C_g at the input.

The transfer function can be found with equation (4.35):

$$T_g(s) = \frac{\tilde{V}_i}{\tilde{V}_{\text{sig}}} = \frac{R_1 \| R_2}{R_{\text{sig}} + R_1 \| R_2} \frac{s\left(R_{\text{sig}} + R_1 \| R_2\right)C_g}{1 + s\left(R_{\text{sig}} + R_1 \| R_2\right)C_g} \approx \frac{s}{s + \omega_g}. \qquad (4.39)$$

For a standard signal source, $R_{\text{sig}} = 50\,\Omega$. On the other hand, R_1 and R_2 are usually in the megaohm range, so R_{sig} is negligible. The corner frequency of this capacitance is at $\omega_g = 1/[(R_1 \| R_2)C_g]$ or $f_g = 1/[2\pi(R_1 \| R_2)C_g]$ In the design process one needs to select the value of C_g. For example, suppose $f_g = 100$ Hz and $R_1 = R_2 = 2\,\text{M}\Omega$, and then one obtains $C_g = 5/[2\pi(R_1 \| R_2)f_g] \approx 8$ nF. A factor of 5 is included so that the -3 dB loss should be avoided.

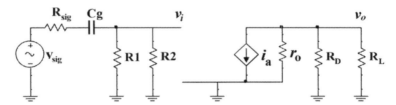

Figure 4.17. Small signal circuit with C_g.

4.5.2 Drain capacitance

Figure 4.18(a) shows the small signal circuit with C_g and C_s shorted. With Thévenin's theorem it can be transformed into a familiar circuit, which is shown in

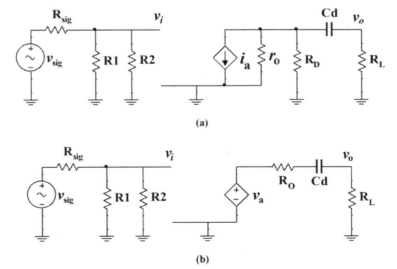

(a)

(b)

Figure 4.18. (a) Small signal circuit with C_d and (b) its equivalent circuit.

figure 4.18(b). First, r_o and R_D can be combined into the output resistance $R_o = r_o \| R_D$. Second, the current source can be converted into a voltage source $v_a = i_a R_o$.

The transfer function of the output section can be found with equation (4.35):

$$T_d(s) = \frac{\tilde{V}_o}{\tilde{V}_a} = \frac{R_L}{R_o + R_L} \frac{s}{s + \omega_d}. \tag{4.40}$$

The corner frequency of this capacitance is at $\omega_d = 1/[(R_o + R_L)C_d]$ or $f_d = 1/[2\pi(R_o + R_L)C_d]$. In the design process one needs to select the values of C_d. For example, suppose $f_d = 100\,\text{Hz}$ and $R_o = R_L = 5\,\text{k}\Omega$, then $C_d = 5/[2\pi(R_o + R_L)f_d] \approx 800\,\text{nF}$. Because R_o and R_L are much lower than R_1 and R_2, C_d needs to be much higher than C_g.

4.5.3 Source capacitance

The analysis of the source capacitance is a little more complicated, so the simplified T-model is used here and the resulting small signal circuit is shown in figure 4.19.

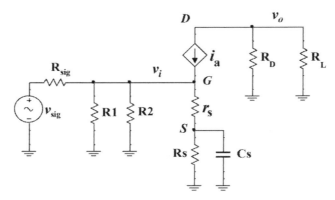

Figure 4.19. Small signal circuit with C_s.

The frequency response is concentrated in the section of the circuit at the bottom of the T-model, where the input signal is v_i at the gate and the output signal is v_{gs} across the resistor r_s. If the positions of r_s and the RC parallel circuit below it are swapped, then it will become the RCR step-up filter circuit shown in figure 4.14(a), and the frequency response is described in equation (4.38).

$$T_s(s) = \frac{\tilde{V}_{gs}}{\tilde{V}_i} = \frac{s + \omega_z}{s + \omega_s}. \tag{4.41}$$

In this equation $\omega_z = 1/R_s C_s$ and $\omega_s = 1/(R_s \| r_s)C_s$. Usually r_s is smaller than R_s, so the transfer function below ω_z becomes quite low: $T(\omega \ll \omega_z) = \omega_z/\omega_s = r_s/(R_s + r_s)$, and this can bring about a significant drop in the voltage gain. Suppose $f_s = 100\,\text{Hz}$ and $g_m = 10\,\text{mS}$, then the required C_s can be estimated: $C_s \approx 5/(r_s \omega_s) = 5g_m/(2\pi f_s)$, and it is about $80\,\mu\text{F}$. Here a factor of 5 is used again so that the deduction of the voltage gain at f_s can be avoided. By comparing the values of these three capacitors, one can find the following relationship: $C_s \gg C_d \gg C_g$.

Suppose these three poles are widely separated from each other so that they can be analyzed independently, then the overall voltage of the CS amplifier can be found as

$$A_V(s) = -g_m(r_o \| R_D)T(s). \tag{4.42}$$

The frequency response is the product of the three transfer functions:

$$T(s) = T_g T_d T_s = \frac{R_L}{r_o \| R_D + R_L} \frac{s}{s + \omega_g} \frac{s}{s + \omega_d} \frac{s + \omega_z}{s + \omega_s}. \tag{4.43}$$

This function determines the shape of the gain profile at low frequency shown in figure 4.15.

4.6 High frequency response of CS amplifier

At high frequency the voltage gain of all amplifiers will drop off at a certain point; a generic profile is shown in figure 4.15. The major cause of this limitation is the effect of the internal capacitors, and a small signal model of a MOSFET with these internal capacitors can be found in figure 4.5.

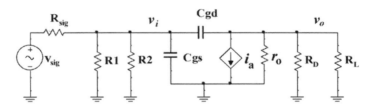

Figure 4.20. High frequency small signal circuit of CS amplifier.

The high frequency response can be analyzed by inserting this model into the AC circuit of the CS amplifier, while shorting out all the external capacitors. One challenge in a circuit such as shown in figure 4.20 is the capacitor C_{gd} on the feedback path between the gate and drain nodes. In section 3.6 we learned how to deal with such a feedback impedance problem, and it can be split up into two parts and grounded separately. The resulting circuit is shown in figure 4.21.

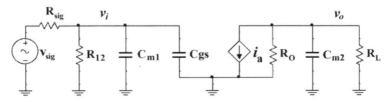

Figure 4.21. Transformed small signal circuit of CS amplifier.

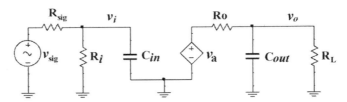

Figure 4.22. Equivalent small signal circuit of CS amplifier.

With Miller's theorem these two equivalent capacitors can be found with equation (3.37):

$$C_{m1} = (K + 1)C_{gd}, \qquad C_{m2} = \frac{K + 1}{K}C_{gd}. \qquad (4.44)$$

The magnitude of the internal gain K can be found at midband:

$$K = g_m(r_o \| R_D \| R_L). \qquad (4.45)$$

As long as K is large enough, $C_{out} = C_{m2} \approx C_{gd}$ is a good approximation. Now we can combine the two parallel capacitors on the input side together:

$$C_{in} = C_{m1} + C_{gs} = (K + 1)C_{gd} + C_{gs}. \qquad (4.46)$$

Similarly, the parallel resistors can also be combined: $R_i = R_1 \| R_2$ and $R_o = r_o \| R_D$. Next, the current source can be converted into a voltage source, and the resulting circuit is shown in figure 4.22.

This circuit can be divided into three sections. At the core is the familiar intrinsic voltage gain:

$$A_{VO} = \frac{\tilde{V}_a}{\tilde{V}_i} = -g_m R_o. \qquad (4.47)$$

Both the input and the output sections have the form of an RCR LPF, so their transfer function can be found from equation (4.34):

$$T_{in}(s) = \frac{\tilde{V}_i}{\tilde{V}_{sig}} = \frac{R_i}{R_{sig} + R_i}\frac{\omega_{in}}{s + \omega_{in}} \approx \frac{\omega_{in}}{s + \omega_{in}}. \qquad (4.48)$$

$$T_{out}(s) = \frac{\tilde{V}_o}{\tilde{V}_a} = \frac{R_L}{R_o + R_L}\frac{\omega_{out}}{s + \omega_{out}}. \qquad (4.49)$$

The pole frequencies are $\omega_{in} = 1/(R_{sig} \| R_{in})C_{in}$ and $\omega_{out} = 1/(R_o \| R_L)C_{out}$, and the lower one is the dominant pole. With these formulae the voltage gain with its high frequency response can be found:

$$A_V(s) = T_{in}(s)A_{VO}T_{out}(s) = -g_m(r_o \| R_D \| R_L)\frac{\omega_{in}}{s + \omega_{in}}\frac{\omega_{out}}{s + \omega_{out}}. \qquad (4.50)$$

Actually, the voltage gain can be derived directly from the original circuit shown in figure 4.20 without using Miller's theorem, and the result is slightly different:

$$A_V(s) = -g_m(r_o \| R_D \| R_L) \frac{\omega_{in}}{s + \omega_{in}} \frac{\omega_{out}}{s + \omega_{out}} (1 - s/\omega_z^*). \tag{4.51}$$

There is an extra factor $(1-s/\omega_z^*)$, where $\omega_z^* = g_m/C_{gd}$, which is usually much higher than ω_1 and ω_2. Therefore, this factor has little effect on the high frequency response below the corner frequency.

For completeness, the low frequency response can also be included, and the complete voltage gain is

$$A_V(s) = F_L(s) A_{VM} F_H(s) = \left(\frac{s}{s + \omega_g} \frac{s}{s + \omega_d} \frac{s + \omega_z}{s + \omega_s} \right)$$

$$[-g_m(r_o \| R_D \| R_L)] \left(\frac{\omega_{in}}{s + \omega_{in}} \frac{\omega_{out}}{s + \omega_{out}} \right). \tag{4.52}$$

This equation generates the frequency response profile shown in figure 4.15.

IOP Concise Physics

The Tao of Microelectronics

Yumin Zhang

Chapter 5

Differential amplifiers

The configurations of amplifiers discussed in chapters 3 and 4 are used in discrete circuits, where large coupling capacitors are available. In ICs, the differential amplifier (DA) is the configuration of choice, since no such capacitors are needed. The history of the DA can be traced back to the vacuum tube age, and the first patent on it was filed by Alan Blumlein in 1936. The idea of the DA originated from the bridge circuit, where the two lower resistors were replaced by two vacuum tubes or transistors. One of the salient advantages of the DA is its tolerance to variations in device characteristics, as long as these characteristics are the same in a pair of vacuum tubes or transistors. In the IC fabrication process, the variations in dimensions and doping levels of the transistors from the specified values are relatively large, but the mismatch of these properties between two transistors next to each other on the same chip is very small, and thus the DA is the ideal configuration for ICs.

5.1 Ideal differential amplifiers

Figure 5.1 shows an ideal DA implemented with BJTs, where an ideal current source is at the bottom. The two input ports are from the bases, and the two output ports are from the collectors. The voltage source VEE at the bottom has a negative bias voltage. In this way, the DC level at the input ports can be zero, so that a signal without a DC offset can be connected directly. The BJTs can be replaced by MOSFETs or vacuum tubes, but the configuration is the same. If the two inputs are identical, then the circuit is in balance and the two output signals are also the same. On the other hand, if the two inputs are different, this difference will be amplified at the output ports.

The operating principle of the DA is analogous to the bipartisan competition in political systems: the current source at the bottom plays the role of total votes from the voters, which are divided by the two branches—equivalent to two political

© Morgan & Claypool Publishers 2014

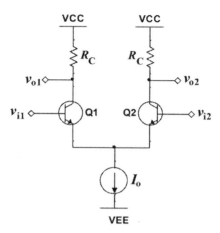

Figure 5.1. Structure of a differential amplifier with an ideal current source.

parties. If one party has some advantage over its rival party, such as more campaign funding, it can get more than half of the total votes and the other party receives the remainder. For the DA circuit shown in figure 5.1, the relationship between the emitter currents and input voltages can be found in the following equations:

$$i_{E1} = \frac{I_o}{1 + \exp\left(-\dfrac{v_{i1} - v_{i2}}{V_T}\right)}, \qquad i_{E2} = \frac{I_o}{1 + \exp\left(\dfrac{v_{i1} - v_{i2}}{V_T}\right)}. \qquad (5.1)$$

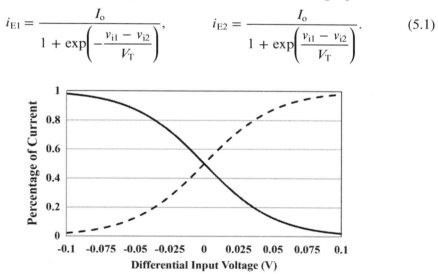

Figure 5.2. I–V characteristics of a differential amplifier.

Figure 5.2 shows the current distribution in the two branches, and the dashed/solid curve represents the current in the left/right branch, respectively. The horizontal axis is the difference between the two input voltages, which is also called the differential input: $v_{id} = v_{i1} - v_{i2}$. It can be verified easily: the sum of these two currents is always equal to the bias current, $i_{E1} + i_{E2} = I_o$. If the differential input voltage is very small, there is a linear region in the I–V curve for analog mode operation that can be well described by the transconductance of transistors. For a BJT DA, this region is within V_T, the thermal voltage.

On the other hand, if this differential input voltage is pretty large, the DA will work in the digital mode and the total current will be swept into one side. Actually, it does not need too much differential input voltage to achieve this. For a BJT DA, 88% of the total current will go to one branch when $v_{id} = 2V_T$. Such a digital circuit configuration is called *emitter-coupled logic* (ECL), and its switching speed is much higher than the traditional BJT digital circuits, since the transistors remain in active mode during switching in ECL and can avoid the delay caused by charge accumulation in the base region. The early models of Cray supercomputers engaged ECL to achieve high speed in computation, but the drawback is the constant current and resulting high power consumption. In order to remove the large amount of heat generated, the circuit boards had to be placed on copper plates with circulating water as the cooling agent.

In general, the input signal on each side can be partitioned into a *common-mode* and a *differential-mode* input. For example, if the input voltages are $v_{i1} = 1.005$ V and $v_{i2} = 0.995$ V, the common-mode input is their average, $v_{icm} = 1$ V, and the differential mode input is their difference, $v_{id} = 0.01$ V. With these definitions, the input signals can be expressed in this way:

$$v_{i1} = v_{icm} + \frac{1}{2}v_{id}, \qquad v_{i2} = v_{icm} - \frac{1}{2}v_{id}. \tag{5.2}$$

In most applications, both the common-mode and the differential-mode inputs are dynamic signals, but the responses to them from the DA are very different. For an ideal DA, shown in figure 5.1, there is no response to the common-mode input at all. In order to simplify the analysis, the differential-mode input can be turned off: $v_{i1}(t) = v_{i2}(t) = v_{icm}(t)$. Although this input signal is changing with time, it cannot break the symmetry of the current, and thus the output voltages at the collector nodes remain constant: $v_{O1} = v_{O2} = V_{CC} - \frac{1}{2}\alpha I_O R_C$.

This is an important figure of merit for DA, which is called *common-mode rejection*. For example, suppose the input signals come from a sensor via two long wires, which can pick up a lot of noisy signals from the environment with magnitudes much higher than the weak differential signal from the sensor. Fortunately, these interference signals are in the common mode and will eventually be rejected by the DA. As a result, only the differential-mode signals from the sensor are amplified. Although the voltage at the collector nodes of the idea DA is not affected by the common-mode input, the voltage at the joint emitter node does follows the change: it is just like two emitter followers joined together.

In the analysis of the differential-mode input, we can turn the common-mode input off for simplification. In this case, the voltage of the emitter node can be considered constant, and it becomes AC grounded in small signal analysis. Therefore, each side of the DA becomes a CE amplifier and the output signals can be derived:

$$v_{o1} = -\frac{1}{2}g_m v_{id} R_C, \qquad v_{o2} = \frac{1}{2}g_m v_{id} R_C. \tag{5.3}$$

The difference of these two output signals is called the differential output:

$$v_{od} = v_{o2} - v_{o1} = g_m v_{id} R_C. \qquad (5.4)$$

With these equations the formulae of single-ended and differential output voltage gains can be derived:

$$A_d = \frac{v_{o2}}{v_{id}} = \frac{1}{2} g_m R_C, \qquad A_{dd} = \frac{v_{od}}{v_{id}} = \frac{v_{o2} - v_{o1}}{v_{id}} = g_m R_C. \qquad (5.5)$$

5.2 Basic differential amplifiers

The DA circuit shown in figure 5.1 requires an ideal current source. Unfortunately, this is unfeasible in practice. A simple implementation is to replace it with a resistor, but the performance is rather poor. A better replacement is a transistor, which functions as a combination of a current source and a resistor. In the next section, more advanced current sources will be discussed and the effect of the resistance can be minimized.

5.2.1 Cheap differential amplifiers

Figure 5.3(a) shows a *cheap* DA circuit, where the current source is replaced by a resistor R_E. First, it works well in the differential mode, so equation (5.5) is still valid. Second, the complete common-mode rejection is lost. As discussed in the previous section, the voltage at the joint emitter node follows the common-mode input signal, which causes a change in the total current going through R_E in this circuit. In the language of the analogous political system, the total number of votes changes with the

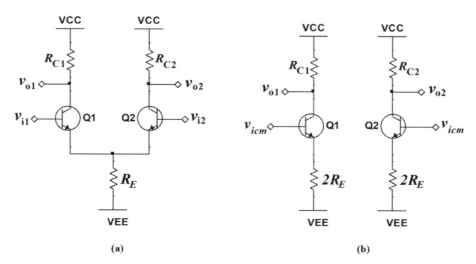

Figure 5.3. (*a*) Cheap differential amplifier and (*b*) its half circuits in common mode.

campaign effort. As a result, the collector current will also change with the common-mode input, and it in turn causes the variation of the output signal at the collector node.

If the differential-mode input is turned off, the symmetry of this circuit allows us to split it into two identical half circuits, which is shown in figure 5.3(b). As the two emitter degenerate resistors in parallel should be identical to the original R_E, the value of this resistance in each branch must be twice that of R_E. Each half of the circuit is equivalent to an amplifier with emitter degenerate resistor shown in figure 3.22, so the formula of the voltage gain in equation (3.39) can be used here:

$$A_{cm} = \frac{v_{o1}}{v_{icm}} = \frac{v_{o2}}{v_{icm}} = -\frac{g_m R_C}{1 + 2g_m R_E}. \tag{5.6}$$

Keep in mind that this common-mode gain is harmful, and thus the lower the better. Therefore, a large R_E seems to be more desirable. However, $2g_m R_E = \alpha(V_E - V_{EE})$ and $V_E \approx -0.7\,V$, so there is little improvement in this configuration. In addition, the design of a DA needs to take into account both the differential-mode and common-mode signals, and the output signal is the superposition of these two components:

$$v_o(t) = A_d v_{id}(t) + A_{cm} v_{icm}(t). \tag{5.7}$$

In many applications, the weak differential-mode signal needs to be singled out from the stronger common-mode interference, so an important figure of merit of DA is defined as the *common-mode rejection ratio* (CMRR), which is the ratio of the magnitudes of the gains in these two modes:

$$\text{CMRR} = 20 \log_{10} \left| \frac{A_d}{A_{cm}} \right| = 20 \log_{10}\left(0.5 + g_m R_E\right). \tag{5.8}$$

In this kind of DA there is little room to improve CMRR, since the transconductance and the emitter resistance are strongly correlated: $g_m R_E = 0.5\alpha(|V_{EE}| - 0.7)/V_T$. It is unrealistic to suggest increasing $|V_{EE}|$, which will cause higher power consumption. Fortunately, the common-mode signal can be eliminated by taking the differential output:

$$v_{od,cm} = v_{o2,cm} - v_{o1,cm} = \frac{g_{m1} R_{C1} - g_{m2} R_{C2}}{1 + 2\bar{g}_m R_E} v_{icm}. \tag{5.9}$$

If the pairs of transistors and collector resistors are well matched ($g_{m1} \approx g_{m2}$ and $R_{C1} \approx R_{C2}$), this gain of differential-mode output with common-mode input becomes very small, so the interference signal can still be rejected. In very sensitive measurements this approach is widely adopted.

5.2.2 Simple differential amplifiers

In order to increase the CMRR, both the transconductance g_m and the emitter resistor R_E must be high. This can be achieved by replacing R_E with a transistor, which is shown in figure 5.4(a). Here the transistor is effectively a combination of a current source I_o and an output resistance R_{om}; the equivalent circuit is shown in figure 5.4(b).

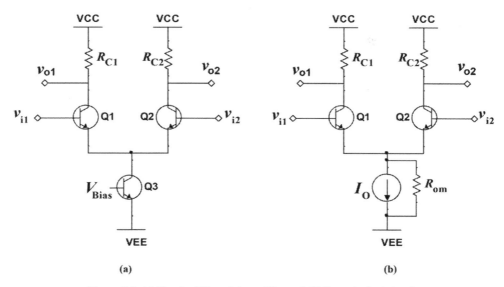

Figure 5.4. (*a*) Simple differential amplifier and (*b*) its equivalent circuit.

This can be understood directly from the I–V characteristics of transistors in active mode, which is shown in figure 5.5. The I–V curve of an ideal current source is a horizontal line, as the current is independent of the voltage across it. However, the I–V characteristics of transistors are not perfectly flat, and the variation of collector or drain voltage will cause a small change in the current. In order to model this *imperfectness*, a resistor R_{om} is attached to the current source in parallel, which reflects the slope of the I–V curves. If a single transistor is used, as in the circuit shown in figure 5.4(*a*), the output resistance of the hybrid-π model plays this role: $R_{om} = r_o$. However, in the next section we will discuss more advanced configurations, where R_{om} can be much higher than r_o.

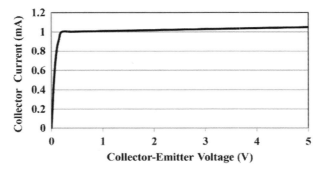

Figure 5.5. I–V curve of transistor.

The analysis of differential-mode and common-mode gains can be separated. In the differential-mode analysis, the output resistance R_{om} can be ignored and the formulae of the gain for single-ended output and differential output are the same as those in equation (5.5). In the case of common-mode input, the current source I_o can be ignored, so it becomes the circuit shown in figure 5.3. With the same approach, the common-mode gain can be found with the formula in equation (5.6), as well as the CMRR. The transconductance of the transistor pair is $g_m = \alpha I_o/(2V_T)$, and the output resistance of the tail transistor is $R_{om} = r_o = V_A/I_o$, so the CMRR can be found:

$$\text{CMRR} = 20\log_{10}\left|\frac{A_d}{A_{cm}}\right| = 20\log_{10}\left(0.5 + g_m R_{om}\right)$$

$$= 20\log_{10}\frac{1 + \alpha V_A/V_T}{2} \approx 20\log_{10}\left(\frac{V_A}{2V_T}\right). \tag{5.10}$$

This CMRR is pretty high, for example $V_A = 100$ V, $V_T = 25.9$ mV, and CMRR = 66 dB. In order to increase CMRR further, either the transconductance g_m or the output resistance R_{om} should be increased. However, with the limitation of power consumption, g_m must remain relatively low, so the only option is to increase R_{om}. In the next section, various current mirror circuits with high R_{om} will be discussed.

5.3 Current mirrors

Discrete amplifier circuits are biased by voltage sources, while integrated amplifier circuits are driven by current sources, though voltage sources are also needed. Therefore, a number of current sources should be created for multiple state amplifier circuits, and this is achieved by *current mirrors*. Simply speaking, a reference current can be set up first and then it is *mirrored* by several other current sources. By changing the size of the transistors serving as these current sources, the mirrored currents can be different from the reference current.

5.3.1 Basic current mirror

Figure 5.6 shows a basic current mirror circuit with BJTs and MOSFETs. We will analyze the BJT circuit in detail, and the MOSFET current mirror can be analyzed in a similar way. The left-hand side of the BJT current mirror is the reference circuit, and the BJT Q1 is in the diode connection configuration, so the collector–emitter voltage is about 0.7 V. With this approximation, the reference current can be designed by choosing R_{ref}:

$$I_{ref} = \frac{V_{CC} - 0.7}{R_{ref}}. \tag{5.11}$$

If the two transistors (Q1 and Q2) are identical and the voltages at the two collector nodes are the same, then the collector currents should be identical: $I_{C2} = I_{C1}$. This current is slightly different from the reference current I_{ref}, which also includes the two base currents: $I_{ref} = I_{C1} + I_{B1} + I_{B2}$. If the BJTs are replaced by MOSFETs,

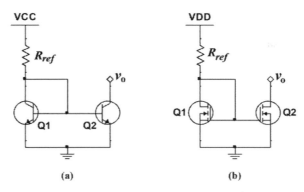

Figure 5.6. Basic current mirror with (*a*) BJTs and (*b*) MOSFETs.

the reference current should be equal to the drain current, since there is no gate current. With simple derivation, the collector current can be found:

$$I_{C2} = \frac{\beta}{\beta + 2} I_{ref}. \tag{5.12}$$

When the voltage at the output node v_O changes, the collector current of Q2 will vary slightly, which can be described by the output resistance r_o:

$$I_O = I_{C2} + (V_O - 0.7)/r_o. \tag{5.13}$$

In this equation, I_{C2} represents a constant current source, and the second term reflects the effect of the parallel output resistance. For the MOSFET current mirror, a similar equation can be derived, but the 0.7 V in the second term should be replaced by the DC gate voltage.

5.3.2 Widlar current mirror

A simple way to boost the output resistance is by adding an emitter degenerate resistor below the emitter of Q2. In order to maintain the symmetry, the same resistor is also placed below the emitter of Q1; the resulting circuit is shown in figure 5.7(*a*). The reference current can also be calculated easily:

$$I_{ref} = \frac{V_{CC} - 0.7}{R_{ref} + R_E}. \tag{5.14}$$

Now we can figure out the output resistance. The small signal circuit of Q2 is shown in figure 5.7(*b*). As an approximation, the base voltage is considered fixed by the reference circuit on the left, so it is an AC ground in the small signal circuit. There are two different approaches to finding the output resistance: applying a test voltage and finding the response current or vice versa. Here, v_o on the right is the test voltage, and we are looking for the current going into the circuit (i_o) from this voltage source. First we can set up an equation with KCL:

$$i_o = i_a + \frac{v_o - v_e}{r_0} = \frac{v_e}{r_\pi} + \frac{v_e}{R_E} = \frac{v_e}{r_\pi \| R_E}. \tag{5.15}$$

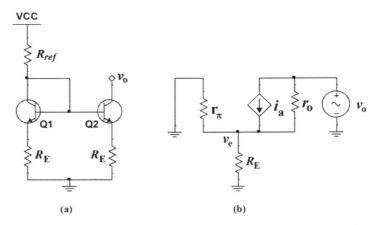

Figure 5.7. (*a*) Widlar current mirror; (*b*) small signal circuit of Q2.

If conductance is used instead of resistance, this equation looks much easier to handle:

$$g_o v_o = \left(g_m + g_o + g_\pi + g_E\right) v_e \rightarrow v_e = \frac{v_o}{1 + g_m r_o + \left(g_\pi + g_E\right) r_o} \approx \frac{v_o}{g_m r_o + r_o/(r_\pi \| R_E)}. \tag{5.16}$$

Plug the expression for v_e into equation (5.15), and the response current can be found:

$$i_o = \frac{v_e}{r_\pi \| R_E} = \frac{v_o}{r_o} \frac{1}{1 + g_m(r_\pi \| R_E)} \rightarrow R_{om} = \frac{v_o}{i_o} = r_o\left[1 + g_m(r_\pi \| R_E)\right]. \tag{5.17}$$

For a Widlar current mirror with **MOSFETs**, this formula can be revised with $r_\pi = \infty$, so the output resistance is

$$R_{om} = r_o\left(1 + g_m R_E\right). \tag{5.18}$$

In section 3.6 we analyzed an amplifier circuit with emitter degenerate resistor from the feedback point view, and this factor $\left(1 + g_m R_E\right)$ is the *amount of feedback*. Widlar current mirror circuits can also be analyzed in this way, and the output resistance of the basic current mirror r_o is enhanced by a factor of the amount of feedback.

5.3.3 Cascode current mirror

In ICs, a resistor is much more expensive than a transistor, as the former occupies much larger chip area. Therefore, it is more economical to replace resistors with transistors, and this often results in better performance as well. In a cascode current mirror, the emitter degenerate resistor R_E in the Widlar current mirror is replaced by a transistor with an output resistance r_o, which is usually much larger than R_E.

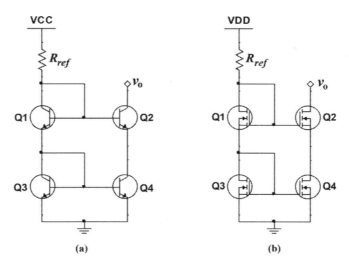

Figure 5.8. Cascode current mirror with (*a*) BJTs and (*b*) MOSFETs.

Figure 5.8 shows the circuit implemented with BJTs and MOSFETs, and the output resistances can be found from equations (5.17) or (5.18).

$$\text{BJT:} \quad R_{\text{om}} \approx r_{o2}\left[1 + g_{m2}(r_{\pi2}\,\|\,r_{04})\right] \approx r_{o2}\left(1 + g_{m2}r_{\pi2}\right) \approx \beta_2 r_{o2}$$

$$\text{MOS:} \quad R_{\text{om}} \approx r_{o2}\left(1 + g_{m2}r_{04}\right) \approx g_{m2}r_{o2}r_{04}.$$

(5.19)

The cascode current mirror is a very popular configuration with high output resistance, especially the **MOSFET** version, but it suffers from one problem: the voltage floor at the output node (v_O) is too high. In modern ICs, V_{CC} or V_{DD} is rather low and still keeps decreasing. In order to get more breathing room, the minimum voltage of v_O needs to be pushed down towards the floor. As we know, the diode connection guarantees the transistor works in the active mode, but it is very conservative. For example, the requirement of active mode for BJTs is $V_{CE} > 0.2$ V or lower, but the diode connection has $V_{CE} = 0.7$ V. Therefore, breaking the diode connection can lower the voltage floor of these current mirror circuits, but it requires an additional subcircuit to generate the base or gate voltages.

5.3.4 Wilson current mirror

The Wilson current mirror is another useful configuration in discrete circuits, but it is not as popular as the cascode current mirror in ICs due to the high voltage floor at the output node. Figure 5.9(*a*) shows a Wilson current mirror with three transistors; a variation of this circuit is shown in figure 5.9(*b*), which is more symmetric but with the same characteristics. The current source at the top of the circuit is actually from another current mirror. In order to simplify the analysis, an ideal current source is used.

The current of a MOSFET is controlled by both v_{GS} and v_{DS}, and their influence can be found in the following equation:

$$\Delta i_D = g_m \Delta v_{GS} + \Delta v_{DS}/r_o = g_m \Delta v_{GS} + g_o \Delta v_{DS}.$$

(5.20)

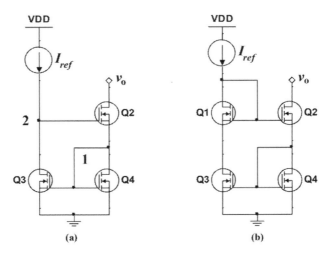

Figure 5.9. Wilson current mirrors.

Since $g_m \gg g_o = 1/r_o$, the current is much more sensitive to the change of v_{GS} than v_{DS}. Therefore, we can imagine a MOSFET as a current source instrument with two control knobs: v_{GS} is the coarse control and v_{DS} is the fine control.

In general, the output resistance of a network can be found in two different ways: applying a test voltage and finding the response current or applying a test current and finding the response voltage. In this circuit the second approach is more appropriate. Suppose there is an increase of the drain current of Q2 (i_O), the drain current in Q4 increases too, as they should be identical. Because the gate and drain of Q4 are tied together, the voltage at this node must increase, and so does the gate voltage of Q3. However, the current of Q3, which is controlled by the current source above it, is constant. The only possibility is that the drain voltage of Q3 is reduced considerably. This in turn will cause a reduction of v_{GS2} of Q2, which incurs a negative feedback to the increase of the output current i_O. This process can be expressed in the following derivation steps. Most of the numbers in the subscripts refer to the transistor numbers, except that v_1 and v_2 stand for the voltages at nodes 1 and 2 in figure 5.9(a), respectively.

$$\Delta i_O \equiv \Delta i_{D2} = g_{m2}\Delta v_{GS2} + g_{o2}\Delta v_{DS2} \approx g_{m2}\Delta v_{GS2} + g_{o2}\Delta v_O \tag{5.21}$$

$$\Delta i_O = \Delta i_{D4} = (g_{m4}+g_{o4})\Delta v_1. \tag{5.22}$$

Combining these two equations results in

$$g_{m2}\Delta v_{GS2} + g_{o2}\Delta v_O = (g_{m4}+g_{o4})\Delta v_1. \tag{5.23}$$

The current of Q3 remains constant:

$$\Delta i_{D3} = g_{m3}\Delta v_1 + g_{o3}\Delta v_2 = 0 \rightarrow \Delta v_2 = -\frac{g_{m3}}{g_{o3}}\Delta v_1 \tag{5.24}$$

$$\Delta v_{GS2} = \Delta v_2 - \Delta v_1 = -\left(1 + \frac{g_{m3}}{g_{o3}}\right)\Delta v_1. \tag{5.25}$$

This equation indicates $|\Delta v_{GS2}| \gg \Delta v_{GS4} = \Delta v_1$ and the change is in the opposite direction. It means that the increase of the output current i_O causes a slight increase of v_{GS4}, but the amount of deduction in v_{GS2} is much larger. This in turn causes a huge increase in v_O to compensate it and results in a very high output resistance. Plug equation (5.25) into equation (5.23); a relationship between Δv_O and Δv_1 can be found:

$$\Delta v_O = \left[r_{o2}/r_{o4} + (g_{m2} + g_{m4})r_{o2} + g_{m2}g_{m3}r_{o2}r_{o3} \right]\Delta v_1 \approx (g_{m2}r_{o2})(g_{m3}r_{o3})\Delta v_1. \qquad (5.26)$$

This equation indicates that Δv_1 has been amplified twice by Q3 and Q2, respectively. Since Δv_1 is much smaller than Δv_O, the approximation in equation (5.21) is valid. Combining equation (5.26) and equation (5.22) and eliminating Δv_1, the output resistance can be found:

$$R_{om} = \frac{\Delta v_O}{\Delta i_O} = \frac{(g_{m2}r_{o2})(g_{m3}r_{o3})}{g_{m4} + g_{o4}} \approx \frac{(g_{m2}r_{o2})(g_{m3}r_{o3})}{g_{m4}} \rightarrow g_m r_o^2 \text{ (for matched MOSFETs).}$$

$$(5.27)$$

If the MOSFETs are replaced by BJTs, the derivation process is a little more complicated due to the base currents, but the idea of feedback is the same. For BJT Wilson current mirror the output resistance is a little lower: $R_{om} \approx 0.5\beta_2 r_{o2}$.

5.4 Differential amplifiers with active load

As we mentioned in the previous section, transistors are cheaper and better than resistors in ICs. Therefore, the load resistors (R_C or R_D) can also be replaced with transistors, which are called active load.

5.4.1 Differential amplifier with single-ended output

Figure 5.10 shows ideal DA circuits with active loads; the BJT and MOSFET versions work in the same way. As the transistor Q3 is in diode connection, only single-ended output is available from the right-hand branch of the circuit. If an output signal is taken from the collector or drain node on the left-hand branch, there is essentially no voltage gain. With the knowledge accumulated in analyzing various amplifiers, the voltage gain can be written in general: $A_{V1} = -g_{m1}(r_{o1} \| R_{out})$, where R_{out} stands for output resistance from the circuit section above Q1. Looking up from the collector or drain of Q1, the diode connection of the transistor Q3 has an output resistance of r_e or $r_s = 1/g_m$, respectively. Therefore, the magnitude of the voltage gain on this branch is close to unity.

Before we go into detail, let us analyze the MOSFET version DA intuitively. Suppose a positive differential input is applied to the gates of Q1 and Q2—the voltage on the left-hand side is higher than that on the right-hand side. More current is steered to the left branch, which causes the gate voltage of Q3 to drop. Keep in

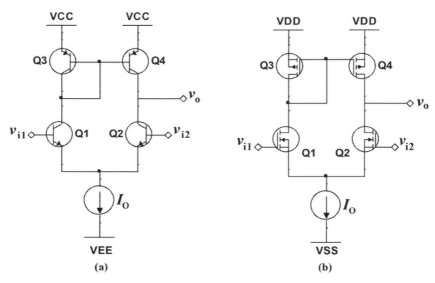

Figure 5.10. DA with active load: (*a*) BJT version and (*b*) MOSFET version.

mind that the magnitude of the drain current is related to the absolute value of the gate–source voltage, $|v_{GS}|$. As the sources of the two *p*-MOSFETs are tied to V_{DD}, lowering the gate voltage contributes to an increase of $|v_{GS}|$ and the drain current.

Due to the diode connection of Q3, its drain voltage decreases by the same amount as the gate voltage. However, being the fine control knob of the current, this small change of drain voltage has little contribution. Unlike the consistent changes on the left branch, there is a conflict on the right branch. The negative input voltage on the gate of Q2 reduces the current, but the lowering gate voltage of Q4 increases the current. Such a conflict has to be resolved by the adjustment of the drain voltage of Q2 and Q4, which play the role of the fine control knobs. Because the influence of the drain voltage to the current is rather weak, it must change a lot to balance the change in the gate voltage, and a high voltage gain is achieved.

Now we can analyze this circuit more rigorously. On the left branch of the MOSFET version DA we can have the following equation for the AC current:

$$i_{\text{left}} \approx g_{m1}\left(\frac{1}{2}v_{\text{id}}\right) \approx -g_{m3}\,v_{g3}. \tag{5.28}$$

In this equation, the contribution of drain voltage change is neglected since it changes very little. In contrast, there is a large change in the drain voltage on the right branch of the circuit, so its contribution must be included.

$$i_{\text{right}} = g_{m2}\left(-\frac{1}{2}v_{\text{id}}\right) + g_{o2}\,v_o = -g_{m4}\,v_{g4} - g_{o4}\,v_o. \tag{5.29}$$

As the gates of Q3 and Q4 are tied together, their voltage is the same: $v_{g3} = v_{g4}$. Suppose the two pairs of transistors are matched: $g_{m1} = g_{m2}$ and $g_{m3} = g_{m4}$, so $g_{m3}v_{g3} = g_{m4}v_{g4}$. Plug equation (5.28) into equation (5.29) and the voltage gain can be found:

$$A_d = \frac{v_o}{v_{id}} = \frac{g_{m2}}{g_{o2} + g_{o4}} = g_{m2}(r_{o2} \| r_{o4}) = g_m(r_{on} \| r_{op}). \tag{5.30}$$

5.4.2 Differential amplifier with differential output

With an external bias voltage for the pair of p-type transistors, a fully differential amplifier can be achieved; the BJT and MOSFET versions are shown in figure 5.11. However, an auxiliary circuit is needed to generate the DC bias voltages V_{B0} and V_{B1} so that the bias current at the bottom is equal to twice the current in each branch. Unlike the pair of n-type transistors, the emitter or source voltages of the p-type transistors are fixed to the power supply V_{CC} or V_{DD}, and a slight change of the DC base or gate bias voltage V_{B1} will cause a large shift of the Q-point of the drain voltage at the output node. Therefore, the design of the DC bias voltages should be consistent.

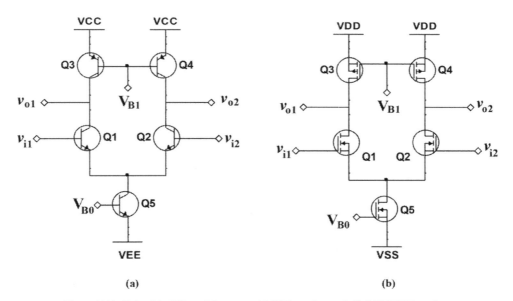

(a)   (b)

Figure 5.11. DA with differential output, (a) BJT version and (b) MOSFET version.

On the other hand, the AC analysis of this circuit is straightforward; the contribution of the pair of p-type transistors is just proving the output resistance r_{op}. With the same approach as used in section 5.1, the voltage gain on each branch can be found:

$$A_{branh} = \frac{v_{o1}}{v_{i1}} = \frac{v_{o2}}{v_{i2}} = -g_m(r_{on} \| r_{op}). \tag{5.31}$$

Written in terms of differential input $v_{i1} = \frac{1}{2}v_d$ and $v_{i2} = -\frac{1}{2}v_d$, the voltage gain is reduced by half:

$$A_d = \frac{v_o}{v_d} = \pm\frac{1}{2}g_m(r_{on}\|r_{op}). \tag{5.32}$$

If the output is taken from the difference of the two output nodes, the fully differential voltage gain is doubled:

$$A_{dd} = \frac{v_{o2} - v_{o1}}{v_d} = g_m(r_{on}\|r_{op}). \tag{5.33}$$

As the output resistance of transistors (r_o) is pretty high, the voltage gain is also very high. For example, for discrete transistors, $r_o \sim 100\,\text{k}\Omega$ and $g_m \sim 20\,\text{mS}$ for $\sim$mA range bias current; the resulting voltage gain is at $|A_{dd}| \sim 1000\,\text{V/V}$. However, in ICs, the output resistance of MOSFETs with short channels is much lower and the current is much weaker, and thus more advanced circuit configurations are needed to achieve high gain. There are two different approaches: vertical and lateral. The vertical approach engages cascode configurations, which can enhance the output resistance as we discussed in the previous section. However, if the bias voltage (V_{CC} or V_{DD}) is rather low, the breathing room of this approach is very limited. The lateral approach refers to multistage amplifiers, which will be covered in the next section.

5.5 Multistage amplifiers

As discussed in the previous section, simple DAs with short channel MOSFETs have the limitation of low voltage gain, and the vertical approach of the cascode configuration has the problem of reduced breathing room in the swing of the output signal. Therefore, the last option is the lateral approach of multistage amplifiers. Besides the high gain, the multistage amplifier also has the advantage of high input resistance and low output resistance, as these parameters can be optimized in separate stages of the circuit. However, the poles in multistage amplifiers are located in a relatively low frequency domain, so an external negative feedback can induce oscillation. The solution to this problem is the introduction of a compensation capacitor, which creates a dominant pole at very low frequency. The drawback of this approach is the reduced bandwidth.

5.5.1 Amplification

Figure 5.12 shows a two-stage DA circuit: at the top is the current mirrors implemented by p-type MOSFETs, in the middle is a DA with active load, and on the right is the second stage amplifier. In order to avoid oscillation, a compensation capacitor C_C is introduced so that it can be amplified by the Miller effect. This circuit uses two power supplies (V_{DD} and V_{SS}); in this way the input signal does not need a DC offset.

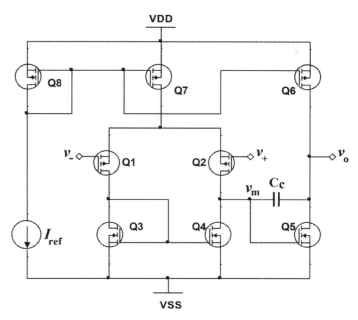

Figure 5.12. Two-stage amplifier circuit.

The analysis of the voltage gains is straightforward; the first stage is basically the same as the circuit shown in figure 5.10. However, the roles of *n*-type and *p*-type MOSFETs are swapped. In addition, the positive input is on the right branch of the DA. The voltage gain of this stage can be found in the same way:

$$A_{V1} = \frac{v_m}{v_{id}} = -g_{m2}(r_{o2} \| r_{o4}). \tag{5.34}$$

The second stage amplification on the right is a simple CS amplifier with an active load, so the gain can be found easily:

$$A_{V2} = \frac{v_o}{v_m} = -g_{m5}(r_{o5} \| r_{o6}). \tag{5.35}$$

The voltage gain of this two stage amplifier is the product of these two expressions:

$$A_V = \frac{v_o}{v_{id}} = \frac{v_o}{v_m} \frac{v_m}{v_{id}} = g_{m2} g_{m5}(r_{o2} \| r_{o4})(r_{o5} \| r_{o6}). \tag{5.36}$$

Suppose $g_m \sim 1$ mS and $r_o \sim 100$ kΩ, the gain is about 68 dB. The gain of this amplifier can be boosted further by adding more stages in amplification, as most operational amplifiers need a gain higher than 100 dB. In addition, this amplifier suffers from a high output resistance, so an output stage is also needed for practical applications.

5.5.2 Frequency response

From the discussions in section 4.5 we know that the low frequency response is due to the external capacitors, i.e. the coupling and bypassing capacitors. However, these capacitors are not present in DA circuits, so they can work well even when the input signal is in DC format.

In the high frequency domain, the stability issue is a serious concern. In the next chapter we will cover the operational amplifier (op-amp), which has a DA at its core. Many op-amp applications involve negative feedback circuits, but the frequency response of the amplifier itself can have a 180° phase shift in a certain frequency domain, and thus an unexpected oscillation can occur. There are two conditions for such an oscillation: the loop gain is greater than unity and the loop phase shift is equal to a multiple of 360°. Let us simplify the frequency response of the amplifier to a two-pole system:

$$A(s) = \frac{A_o}{(1 + s/\omega_{p1})(1 + s/\omega_{p2})}. \tag{5.37}$$

Suppose $\omega_{p2} > \omega_{p1}$ so that ω_{p1} is the dominant pole. In the frequency domain of $\omega > 10\omega_{p2}$, the phase shift becomes 180°.

$$A(\omega > 10\omega_{p2}) \approx \frac{A_o}{\left(j\dfrac{\omega}{\omega_{p1}}\right)\left(j\dfrac{\omega}{\omega_{p2}}\right)} = -\frac{\omega_{p1}\omega_{p2}}{\omega^2}A_o. \tag{5.38}$$

If the loop gain in this frequency domain is greater than unity, negative feedback will become positive feedback and oscillation can start by itself. This has nothing to do with the input signal, and it is an intrinsic problem. In order to prevent such oscillation from happening, a safety margin in phase is usually required when the loop gain drops to unity, which is called the *phase margin*. As a simplified example, we can choose a very aggressive design with the phase margin as low as 45°. At $\omega = \omega_{p2}$ the second factor in the denominator of equation (5.37) gives a phase shift of $-45°$: $1 + s/\omega_{p2} = 1 + j = \sqrt{2}\angle45°$. Suppose $\omega_{p2} > 10\omega_{p1}$ and $\beta = -\beta_o$, then the loop gain at $\omega = \omega_{p2}$ becomes

$$A\beta(\omega = \omega_{p2}) = \frac{A_o\beta_o\angle180°}{\left(\dfrac{\omega_{p2}}{\omega_{p1}}\angle90°\right)\left(\sqrt{2}\angle45°\right)} = \frac{\omega_{p1}}{\omega_{p2}}\frac{A_o\beta_o}{\sqrt{2}}\angle45°. \tag{5.39}$$

If both $A_o\beta_o$ and ω_{p2} are fixed, the stability condition requires

$$|A\beta(\omega = \omega_{p2})| < 1 \rightarrow \frac{\omega_{p1}}{\omega_{p2}} < \frac{\sqrt{2}}{A_o\beta_o}. \tag{5.40}$$

Usually $A_o\beta_o$ is pretty high, so this expression means that the first pole must be much lower than the second pole. For example, if $A_o\beta_o = 10^4$ and $\omega_{p2} = 10^6$ rad/s, then $\omega_{p1} < 100\sqrt{2}$ rad/s. A pole this low usually requires a huge capacitor; in discrete circuits this is not a serious problem, but it is impossible to put it on a silicon chip. In the early history of op-amps, people needed to connect an external capacitor to a port of an op-amp for this purpose, but this is very inconvenient. Fortunately, the Miller effect can be used in amplifying a small capacitor into a much larger one so that it can be included in later designs of op-amps. In the circuit shown in figure 5.12 the capacitor C_C plays this role, and its amplified value becomes

$$C_{m1} = \left[1 + g_{m5}(r_{o5}\|r_{o6})\right]C_C \approx g_{m5}(r_{o5}\|r_{o6})C_C. \qquad (5.41)$$

The output resistance of the DA is $R_{om} = r_{o2}\|r_{o4}$, so the dominant pole frequency is

$$\omega_{p1} = \frac{1}{R_{om}C_{m1}} = \frac{1}{g_{m5}(r_{o5}\|r_{o6})(r_{o2}\|r_{o4})C_C}. \qquad (5.42)$$

Suppose $\omega_{p1} = 100\sqrt{2}$ rad/s, $g_{m5} = 1$ mS, and $r_o \sim 100$ kΩ; then the required capacitance is about $C_C \sim 3$ nF. This value is still too high to be implemented on silicon chips, and it can be further reduced by increasing the output resistance of the DA.

IOP Publishing

The Tao of Microelectronics

Yumin Zhang

Chapter 6

Operational amplifiers

Operational amplifiers (op-amps) are the most widely used analog electronic devices. The name originates from their application in analog computers, since they have the ability to carry out arithmetic operations. Compared with the simple BJT and MOSFET amplifiers covered in chapters 3 and 4, op-amps have much better performance and are also more convenient to use. There are families of op-amps specialized for different applications, but some common characteristics are shared by all of them, such as very high gain, extremely high input impedance, and rather low output impedance.

6.1 Introduction to op-amps

Figure 6.1 shows a functional diagram of an op-amp. At its core is a DA similar to the one shown in figure 5.12. Therefore, there is a simple relationship between the output and input signals:

$$v_o = A_o(v_+ - v_-). \tag{6.1}$$

Most op-amps need two complimentary voltage sources, but there are also op-amps that can be biased by just a single power supply. The range of the output voltage is limited by the rail voltages from the voltage sources (V^+ and V^- in figure 6.1), and usually there is an offset from them. There is also a limitation on the output current, which is determined by the output stage of the op-amps. In applications with a heavy load, power op-amps with a high output current should be selected.

Traditional op-amps have a very low dominant pole, such as the well known LM741 model with its dominant pole at less than 10 Hz. At that time transistors were pretty large, so the frequency of the second pole was not very high. Some

doi:10.1088/978-1-6270-5453-9ch6
© IOP Publishing Ltd 2014

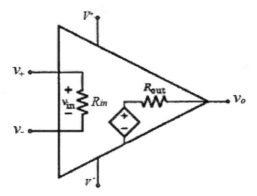

Figure 6.1. Functional diagram of op-amp.

modern op-amps have a much wider bandwidth and the frequency of the dominant pole is pushed towards the megahertz range, since the second pole is increased to the gigahertz range and the DC gain is reduced to 60 dB. As we know, one can always extend bandwidth by reducing the gain, so the figure of merit of an op-amp is the gain–bandwidth product, which is equal to the unit gain frequency: $f_T = A_0 f_{p1}$. A typical op-amp frequency response is shown in figure 6.2. It is the same as a first order LPF with a roll off slope of -20 dB/dec.

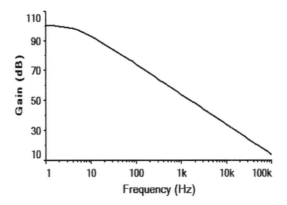

Figure 6.2. Frequency response of op-amp.

A primitive application of op-amps is a *comparator*, where two signals are directly connected to the input ports and the output signal indicates which input signal has a higher voltage. In such a *digital mode* of operation, the output signal jumps between the upper bound of $V_{OH} = V^+ - \Delta V^+$ and the lower bound of $V_{OL} = V^- + \Delta V^-$, where V^+ and V^- stand for the rail voltages while ΔV^+ and ΔV^- refer to the offset voltages. Comparators can be used to digitize analog signals by comparison with a set of discrete values. In addition, they can also be used to trigger an alarm when certain parameters are beyond the safe operating range. This can be considered as the *digital mode* operation.

Most applications of op-amps are in the *analog mode*, where the output signal swings within these two bounds. In order to simplify the analysis some approximations are adopted. First, as the input impedance is extremely high, it is assumed that the currents at the input ports are zero. Second, the gain of op-amps is assumed to be infinite. Thus equation (6.1) requires $v_+ \approx v_-$ so that the output voltage is finite. When one of the input terminals is grounded, the other terminal is considered *virtually grounded*. Third, the slew rate is assumed to be very high and there is essentially no delay between the input and output signals.

6.2 Op-amps with negative feedback

By definition, the primary application of op-amps is the amplification of weak signals. There are two basic circuit configurations: inverting and non-inverting. Both of these configurations involve two resistors forming the feedback path and connecting to the inverting input terminal of the op-amp. The inverting configuration allows many signals to fan in and superpose together, while the non-inverting configuration has high input impedance and the gain is positive.

6.2.1 Non-inverting amplifier

Figure 6.3 shows two equivalent circuit diagrams of a non-inverting amplifier. The input signal is connected to the non-inverting port directly. In the circuit diagram shown in figure 6.3(*b*) we can see more clearly that R_1 and R_2 form a voltage divider circuit. With the relationship of $v_+ \approx v_-$, the input–output relationship can be found:

$$v_i = \frac{R_1}{R_1 + R_2} v_o \rightarrow A_f = \frac{v_o}{v_i} = \frac{R_1 + R_2}{R_1} = 1 + \frac{R_2}{R_1}. \tag{6.2}$$

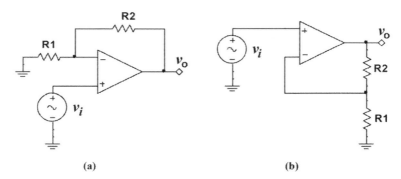

(a)   (b)

Figure 6.3. Non-inverting amplifier.

This closed-loop voltage gain can be derived from the general theory of feedback systems, and the system diagram is shown in figure 6.4.

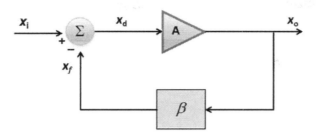

Figure 6.4. System diagram of negative feedback.

There are four variables that can be divided into two groups. The external input signal x_i and the output signal x_o are the two signals outside of the black box, and thus they can be observed and measured directly. Then there are two internal variables inside the black box: the feedback signal x_f and the amplifier input signal x_d. These four variables have the following relationships:

$$x_d = x_i - x_f$$
$$x_o = A x_d \tag{6.3}$$
$$x_f = \beta x_o \quad .$$

With the internal variables (x_f and x_d) represented by the external variables (x_i and x_o), the closed-loop gain can be found:

$$A_f = \frac{x_o}{x_i} = \frac{A}{1 + A\beta}. \tag{6.4}$$

Now we can translate these system variables into circuit variables: $x_i \rightarrow v_+$, $x_o \rightarrow v_o$, $x_f \rightarrow v_-$, and $x_d \rightarrow v_d = v_+ - v_-$. The feedback factor can be found from the voltage divider circuit of R_1 and R_2: $\beta = v_-/v_o = R_1/(R_1 + R_2)$. Since the open-loop gain is very high, the closed-loop gain can be found with an approximation:

$$A_f = \frac{A}{1 + A\beta} \approx \frac{A}{A\beta} = \frac{1}{\beta} = \frac{R_1 + R_2}{R_1} = 1 + \frac{R_2}{R_1}. \tag{6.5}$$

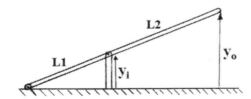

Figure 6.5. Hydraulic crane model of non-inverting amplifier.

In order to understand this formula intuitively, a mechanical model of a hydraulic crane can be developed, which is shown in figure 6.5. The voltages can be converted to heights: $v_i \rightarrow y_i$ and $v_o \rightarrow y_o$. The ground node of the feedback path corresponds to the joint of the hydraulic arm on the left, which remains stationary. In this way the two resistors resemble the lengths of the two sections of the hydraulic arm: $R_1 \to L_1$ and $R_2 \to L_2$. In the operation of the hydraulic crane, L_1 is fixed and L_2 is changeable, but the ratio of the heights and the ratio of the lengths are the same:

$$\frac{y_o}{y_i} = \frac{L_1 + L_2}{L_1} = 1 + \frac{L_2}{L_1}. \tag{6.6}$$

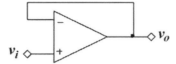

Figure 6.6. Op-amp buffer.

The non-inverting configuration has the advantage of very high input impedance. In addition, the output impedance is very low for both configurations. Therefore, a unit gain *buffer* can be constructed that can be used to isolate two sections of a circuit or as the interface with the external world. Figure 6.6 shows such a buffer circuit, which can be considered as a special form of the non-inverting amplifier with $R_1 = \infty$ and $R_2 = 0$. It has a very deep feedback and the bandwidth is extended to unit gain frequency.

6.2.2 Inverting amplifier

Figure 6.7(a) shows the circuit diagram of an inverting amplifier. In this configuration the input signal is connected to the inverting port of the op-amp through a resistor, but the feedback path is unchanged. With the non-inverting input terminal grounded, the inverting terminal is also grounded virtually, but no current leaks into the op-amp at this node. Therefore, the currents passing through the two resistors in the feedback path are the same.

$$i_f = \frac{v_i - 0}{R_1} = \frac{0 - v_o}{R_2} \to A_f = \frac{v_o}{v_i} = -\frac{R_2}{R_1}. \tag{6.7}$$

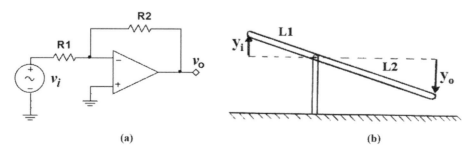

(a)　　　　　　　　　　　　　　　(b)

Figure 6.7. (a) Inverting amplifier and (b) seesaw model.

Just like the intuitive approach to the non-inverting amplifier, a similar mechanical model can also be developed for the inverting amplifier, which is shown in figure 6.7(b). There are things in common: for example, the ground must be a fixed point and the resistance is proportional to the length of a mechanical structure. Usually we have $R_2 > R_1$ so that the magnitude of the gain is greater than unity, but it can also be the other way round. This is not the case for the non-inverting amplifier, as its gain cannot be less than unity.

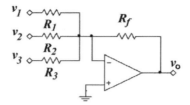

Figure 6.8. Summing amplifier.

A unique feature of inverting amplifier configuration is combining multiple input signals; from the point of view of arithmetic operations, it can carry out the computation of weighted summation. Figure 6.8 shows an example of such a circuit, and it can be analyzed with the same approach as used in the derivation of equation (6.7).

$$i_f = \frac{v_1}{R_1} + \frac{v_2}{R_2} + \frac{v_3}{R_3} = \frac{0 - v_o}{R_f} \rightarrow v_o = -\left(\frac{R_f}{R_1}v_1 + \frac{R_f}{R_2}v_2 + \frac{R_f}{R_3}v_3 \right). \qquad (6.8)$$

6.2.3 Operation of subtraction

As the voltage gains of the inverting and non-inverting amplifiers are of different signs, their combination can perform the operation of subtraction. Figure 6.9 shows a subtracting circuit, which can be analyzed with the superposition principle.

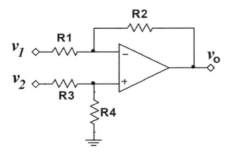

Figure 6.9. Subtracting amplifier.

First, the input signal v_2 is turned off (grounded), and the circuit becomes an inverting amplifier.

$$v_{o1} = -\frac{R_2}{R_1}v_1. \tag{6.9}$$

Next, the input signal v_1 is turned off and we have a non-inverting amplifier.

$$v_{o2} = \frac{R_1 + R_2}{R_1}v_+ = \frac{R_1 + R_2}{R_1}\frac{R_4}{R_3 + R_4}v_2. \tag{6.10}$$

If $R_1 = R_3$ and $R_2 = R_4$, the gain of this non-inverting amplifier will become R_2/R_1. Now we can combine these two equations together:

$$v_o = v_{o1} + v_{o2} = \frac{R_2}{R_1}(v_2 - v_1). \tag{6.11}$$

This result is similar to that of a DA, but the gain can be adjusted easily by changing the values of R_1 and R_2. In practice, buffers are often needed since the input resistance is not very high. An instrumentation amplifier is essentially a subtraction amplifier with buffered inputs, which is widely used in many applications.

6.3 Active filters

In section 4.4 we discussed first order passive RC filters, which can suppress the interference signals in the rejection band but cannot amplify signals in the pass band. Active filters combine RC circuits with op-amps, and thus they can overcome this limitation and boost the signals in the pass band. In addition, compact second order filters can be constructed, and they can reject interference signals close to the pass band more effectively.

6.3.1 First order active filters

Figure 6.10 shows two first order LPF circuits with non-inverting and inverting inputs, respectively. The non-inverting configuration shown in figure 6.10(a) is a combination of an RC LPF and a non-inverting amplifier, so the transfer function reflects these two components:

$$T(s) = \frac{\tilde{V}_o}{\tilde{V}_i} = \left(1 + \frac{R_2}{R_1}\right)\frac{\omega_c}{s + \omega_c}. \tag{6.12}$$

Compared with the non-inverting configuration, the inverting LPF circuit shown in figure 6.10(b) is more compact, and the number of resistors is reduced from

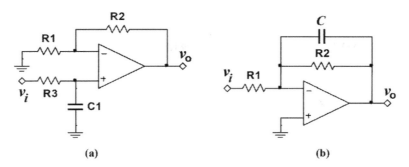

Figure 6.10. First order active low pass filters.

three to two. The transfer function can be derived from the generalized formula of voltage gain:

$$T(s) = \frac{\tilde{V}_o}{\tilde{V}_i} = -\frac{Z_2}{R_1} = -\frac{1}{R_1}\frac{R_2/sC}{R_2 + 1/sC} = -\frac{R_2}{R_1}\frac{\omega_c}{s + \omega_c}. \tag{6.13}$$

In this equation, $\omega_c = 1/(R_2 C)$. The low-pass behavior of this filter can be analyzed intuitively from the two extreme situations. At very low frequencies the capacitor becomes open circuit and this filter is transformed into an inverting amplifier with a gain of $-R_2/R_1$. At very high frequencies the impedance of the capacitor will be much lower than R_2, which is basically bypassed by the capacitor. Therefore, the transfer function becomes $T(s) = -1/(sR_1 C)$.

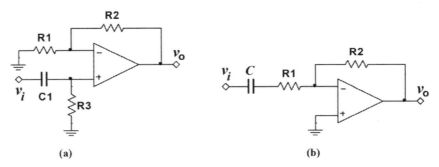

Figure 6.11. First order active high-pass filters.

HPFs can be constructed in a similar way. The non-inverting active HPF shown in figure 6.11(a) is a combination of an RC HPF and a non-inverting amplifier, so the transfer function is the product of the voltage gain and the transfer function of the passive RC HPF:

$$T(s) = \left(1 + \frac{R_2}{R_1}\right)\frac{s}{s + \omega_c}. \tag{6.14}$$

The inverting active HPF shown in figure 6.11(b) is more compact than the non-inverting configuration, and the transfer function can be found with the ratio of the impedances:

$$T(s) = -\frac{R_2}{Z_1} = -\frac{R_2}{R_1 + 1/sC} = -\frac{R_2}{R_1}\frac{s}{s + \omega_c}. \tag{6.15}$$

In this equation, $\omega_c = 1/(R_1C)$. The characteristics of this inverting HPF can also be found from the frequency response of the capacitor. At very high frequencies the capacitor behaves like a short circuit, and this HPF becomes an inverting amplifier. At very low frequencies the impedance of the capacitor is much higher than R_1, which can be ignored under this circumstance. Therefore, the transfer function becomes $T(s) = -R_2/Z_C = -sR_2C$.

6.3.2 Second order active filters

First order filters suffer from the slow roll off away from the pass band. If the interference signals are very close to the selected signal in the frequency domain, such filters cannot suppress the interference very effectively. Sallen–Key filters are a family of popular second order active filters introduced in 1955 by R P Sallen and E L Key from the MIT Lincoln Lab.

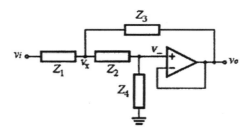

Figure 6.12. Generic Sallen–Key second order active filters.

Figure 6.12 shows the generic Sallen–Key filter; for simplicity in analysis, the inverting input is shorted to the output. If two feedback resistors are connected here, it will have a gain. In this circuit, $v_- = v_+ = v_o$; the only unknown node is $\tilde{V}_x$, which can be figured out from the voltage divider circuit of Z_2 and Z_4:

$$\frac{\tilde{V}_+}{\tilde{V}_x} = \frac{Z_4}{Z_2 + Z_4} \rightarrow \tilde{V}_x = \frac{Z_2 + Z_4}{Z_4}\tilde{V}_o. \tag{6.16}$$

Applying KCL at node Vx, the following equation can be obtained:

$$\frac{\tilde{V}_i - \tilde{V}_x}{Z_1} = \frac{\tilde{V}_x - \tilde{V}_+}{Z_2} + \frac{\tilde{V}_x - \tilde{V}_o}{Z_3}. \tag{6.17}$$

The transfer function can be found by plugging equation (6.16) into equation (6.17):

$$T(s) = \frac{\tilde{V}_o}{\tilde{V}_i} = \frac{Z_3 Z_4}{Z_1 Z_2 + Z_3 (Z_1 + Z_2) + Z_3 Z_4}. \tag{6.18}$$

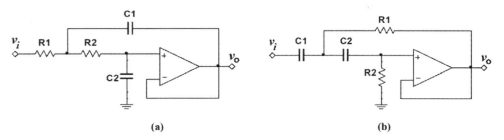

Figure 6.13. Second order active (a) low-pass filter and (b) high-pass filter.

Figure 6.13(a) shows a second order LPF circuit, and the transfer function can be found by replacing the generic impedance with the specific ones in this circuit:

$$T(s) = \frac{\tilde{V}_o}{\tilde{V}_i} = \frac{1}{1 + C_2(R_1 + R_2)s + R_1 R_2 C_1 C_2 s^2} = \frac{\omega_c^2}{s^2 + 2\zeta\omega_c s + \omega_c^2}. \tag{6.19}$$

In this equation, $\omega_c = 1/\sqrt{R_1 R_2 C_1 C_2}$ and $\zeta = 1/[2\omega_c (R_1 \| R_2) C_1]$. At very low and very high frequencies, this transfer function can be simplified:

$$T(s) \approx \frac{\omega_c^2}{\omega_c^2} = 1 \qquad \text{(for } \omega \ll \omega_c\text{)}$$

$$T(s) \approx \frac{\omega_c^2}{s^2} = -\frac{\omega_c^2}{\omega^2} \qquad \text{(for } \omega \gg \omega_c\text{)}. \tag{6.20}$$

In the low frequency domain ($\omega \ll \omega_c$) the capacitors are equivalent to open circuits, so the filter becomes a unit gain buffer. Beyond the corner frequency the Bode plot of the transfer function rolls off at -40 dB/dec. Although the second order filter has a steeper roll off curve, it can resonate around the corner frequency and create a peak in the Bode plot. In order to prevent this from happening, the following condition needs to be satisfied: $\zeta > 1/\sqrt{2}$.

Figure 6.13(b) shows a circuit of a second order active HPF, and its transfer function can also be derived from equation (6.18):

$$T(s) = \frac{\tilde{V}_o}{\tilde{V}_i} = \frac{R_1 R_2 C_1 C_2 s^2}{1 + R_1(C_1 + C_2)s + R_1 R_2 C_1 C_2 s^2} = \frac{s^2}{s^2 + 2\zeta\omega_c s + \omega_c^2}. \tag{6.21}$$

In this equation, $\omega_c = 1/\sqrt{R_1 R_2 C_1 C_1}$ and $\zeta = (C_1 + C_2)/(2\omega_c R_2 C_1 C_2)$. At very low and very high frequencies, this expression can be simplified:

$$T(s) \approx \frac{s^2}{\omega_c^2} = -\frac{\omega^2}{\omega_c^2} \quad \text{(for } \omega \ll \omega_c)$$

$$T(s) \approx 1 \quad \text{(for } \omega \gg \omega_c). \tag{6.22}$$

In the high frequency domain ($\omega \gg \omega_c$), the capacitors can be considered as short circuits, so the filter becomes a unit gain buffer. In the domain below the corner frequency, in the Bode plot the transfer function curve increases at 40 dB/dec, which is much steeper than the first order HPF. In order to avoid ringing, the same condition is required: $\zeta > 1/\sqrt{2}$.

Besides LPFs and HPFs, band-pass filters can also be constructed from the generic Sallen–Key circuit. In addition, band-pass filters can also be implemented by cascading an LPF and an HPF, and signals can pass through the overlapping pass bands of these two filters. Similarly, a band-stop filter can be constructed from an LPF and an HPF in parallel, and signals falling into the gap of the two pass bands will be rejected.

6.4 Op-amps with positive feedback

As we discussed in section 6.2, in either inverting or non-inverting amplifier configurations, the feedback path consisting of two resistors is connected to the inverting input terminal of the op-amp, which provides a negative feedback. If this feedback path is connected to the non-inverting input terminal, the resulting circuit with positive feedback is called a Schmitt trigger, and the output signal will be at either the upper bound ($V_{OH} = V^+ - \Delta V^+$) or the lower bound ($V_{OL} = V^- + \Delta V^-$) of the op-amp.

A system with negative feedback can be imagined as a bowl with a small ball in it; when the ball is pulled away from its lowest position at the center and released, it will go back to the equilibrium position and stay there. In contrast, a system with positive feedback is like a small ball at the top of a flipped bowl; when the ball is pushed away just a little bit, it will roll down the side.

6.4.1 Inverting Schmitt trigger

Figure 6.14(*a*) shows the circuit of an inverting Schmitt trigger, and its behavior is shown in figure 6.14(*b*). This circuit can be considered as a *comparator*, and the feedback path—a voltage divider—sets up two threshold voltages:

$$V_{TL} = \frac{R_1}{R_1 + R_2} V_{OL}, \qquad V_{TH} = \frac{R_1}{R_1 + R_2} V_{OH}. \tag{6.23}$$

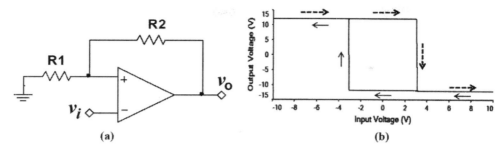

(a)   (b)

Figure 6.14. Inverting Schmitt trigger.

When the input voltage is between these two threshold values, there are two possible output voltages. In this case, the history of the input voltage determines which threshold voltage is active. For example, if the input voltage starts at a value below V_{TL}, the output voltage will be at V_{OH} since the non-inverting input is higher than the inverting input ($v_+ > v_-$). In this case, the feedback path sets the threshold voltage at V_{TH}. Therefore, the output voltage will remain at V_{OH} as the input voltage increases until it reaches this threshold voltage; further increase of the input voltage will toggle the output voltage from V_{OH} to V_{OL}. At the same time, the threshold voltage becomes V_{TL}, which is the threshold voltage when the input voltage sweeps back from a voltage higher than V_{TH}. In this way, a hysteresis loop is created in a round trip sweep of the input voltage.

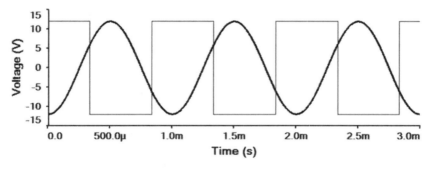

Figure 6.15. Application of Schmitt trigger.

Schmitt triggers are widely used in digitizing analog signals, as this hysteresis path can remove the interference of noise. An example is shown in figure 6.15, and the threshold voltages can be found at the intersection points between these two waveforms. When the input signal increases from below the lower threshold voltage, the output voltage is at the upper bound and the higher threshold voltage is active. On the other hand, when the signal gets above the higher threshold voltage, the output jumps down to the lower bound and the lower threshold voltage becomes active. The drawback of the inverting Schmitt trigger is the reversed polarity.

In many applications the input signal is always positive or negative, but it changes between two different voltage levels, such as a sine wave with a DC offset. In order to deal with this situation, the two threshold voltages should be adjustable.

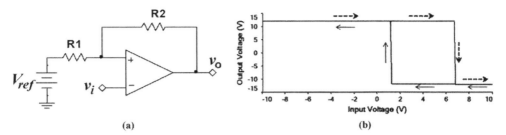

(a)   **(b)**

Figure 6.16. Inverting Schmitt trigger with offset.

Figure 6.16(a) shows such a circuit with a reference voltage connected to the feedback path, and its behavior is shown in figure 6.16(b). The threshold voltages can be derived with the superposition principle:

$$V_{TL} = \frac{R_1 V_{OL} + R_2 V_{ref}}{R_1 + R_2}, \qquad V_{TH} = \frac{R_1 V_{OH} + R_2 V_{ref}}{R_1 + R_2}. \qquad (6.24)$$

The introduction of V_{ref} can shift the center of the two threshold voltages, but the difference between them remains unchanged: $V_{TH} - V_{TL} = [R_1/(R_1 + R_2)]$ $(V_{OH} - V_{OL})$.

6.4.2 Non-inverting Schmitt trigger

The inverting Schmitt trigger has very high input impedance, but the output signal is of the opposite polarity. Figure 6.17(a) shows a non-inverting Schmitt trigger circuit, and its behavior is shown in figure 6.17(b). The advantage of this configuration is that the polarities of the input and output signals are the same, but the input impedance is not very high: $R_{in} = R_1$.

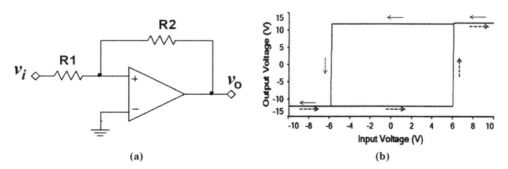

(a)   **(b)**

Figure 6.17. Non-inverting Schmitt trigger.

This circuit can also be analyzed from the point of view of a comparator. From the superposition principle, the voltage at the non-inverting input port can be found:

$$v_+ = \frac{R_1 v_o + R_2 v_i}{R_1 + R_2}. \qquad (6.25)$$

The threshold voltages can be derived with the condition $v_+ = v_- = 0$:

$$V_{TL} = -\frac{R_1}{R_2} V_{OH}, \qquad V_{TH} = -\frac{R_1}{R_2} V_{OL}. \qquad (6.26)$$

If a shift of these threshold voltages is needed, a reference voltage can be applied at the inverting input port. In a similar way, the offset threshold voltages can be derived with the condition $v_+ = v_- = V_{ref}$:

$$V_{TL} = \frac{(R_1 + R_2) V_{ref} - R_1 V_{OH}}{R_2}, \qquad V_{TH} = \frac{(R_1 + R_2) V_{ref} - R_1 V_{OL}}{R_2}. \qquad (6.27)$$

6.5 Oscillators

Figure 6.18 shows a diagram of a positive feedback system, which consists of an amplifier and a feedback path. If the feedback factor β is a function of frequency, such an op-amp circuit can become an oscillator.

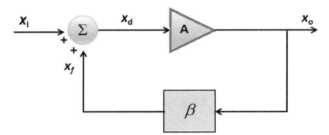

Figure 6.18. System diagram of positive feedback.

Similar to the analysis of the negative feedback system in section 6.2, the relationships of the four internal and external variables can be found from this system diagram:

$$x_d = x_i + x_f$$
$$x_o = A x_d \qquad (6.28)$$
$$x_f = \beta x_o.$$

By replacing the internal variables (x_d, x_f) with the external variables (x_i, x_o), the feedback gain can be derived:

$$A_f = \frac{x_o}{x_i} = \frac{A}{1 - A\beta}. \qquad (6.29)$$

The condition for a stable oscillation—the Barkhausen criterion—is that the denominator vanishes and the feedback gain goes to infinity:

$$A\beta = 1 \rightarrow |A| \cdot |\beta| = 1 \qquad \text{and} \qquad \angle A + \angle \beta = n \cdot 360°. \qquad (6.30)$$

In an oscillator circuit the feedback gain is actually not a good starting point, since there is no input signal. A more relevant expression is the loop gain: $L(s) = A(s)\beta(s)$. The signal can be imagined as a car running around a closed race track, and it is amplified by the loop gain after each round trip. A stable oscillation with a fixed period is called a *limit cycle* in nonlinear dynamic theory, and cannot be implemented with linear circuit elements. In other words, the loop gain cannot be a constant; otherwise, the oscillation cannot start and then be stabilized. Therefore, the loop gain must be a function of the magnitude of the signal. When the signal is very weak, the loop gain is greater than unity so the signal can grow in magnitude. However, when the signal is too strong, the loop gain becomes less than unity so it will get weaker. In this way the magnitude of the signal is constrained to the boundary between these two domains, and this equilibrium point is stipulated by the Barkhausen criterion.

Amplifiers are easier to understand, because there is a clear relationship between input and output signals. On the other hand, oscillators are somewhat mysterious, and it seems a signal can be created from nothing. Actually, the generated signal originates from random noise, which can be considered as a weak broadband signal. If the voltage gain profile of the amplifier is flat on the frequency spectrum, the job of frequency selection falls on the feedback network. The Wien-bridge oscillator is an excellent example to illustrate this mechanism.

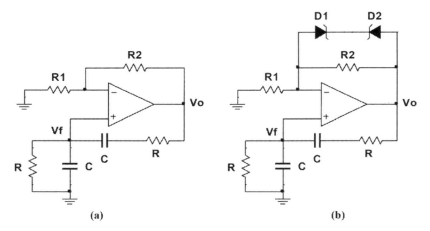

Figure 6.19. Wien-bridge oscillators.

Figure 6.19(a) shows the ideal Wien-bridge oscillator circuit. The inverting input port is connected to a feedback path consisting of two resistors (R_1 and R_2), which determine the voltage gain: $A = 1 + R_2/R_1$. The non-inverting input node is connected to another feedback network consisting of resistors and capacitors, which form the so-called *lead–lag* filter that determines the feedback factor β:

$$\beta(s) = \frac{\tilde{V}_{\mathrm{f}}}{\tilde{V}_{\mathrm{o}}} = \frac{Z_{\mathrm{P}}}{Z_{\mathrm{S}} + Z_{\mathrm{P}}} = \frac{1}{1 + Z_{\mathrm{S}} Y_{\mathrm{P}}}$$

$$= \frac{1}{1 + (R + 1/sC)(1/R + sC)} = \frac{1}{3 + (s/\omega_{\mathrm{o}} + \omega_{\mathrm{o}}/s)}. \tag{6.31}$$

In this equation, Z_S and Z_P are the impedances of the series and parallel RC circuits, and $\omega_o = 1/RC$. The frequency response of this feedback factor is shown in figure 6.20. At the resonance frequency $f_o = 1\,kHz$, the phase shift is zero and the magnitude is at the peak of 1/3. Following the Barkhausen criterion, the voltage gain required is 3, but slightly higher gain is needed to get the oscillation started. In figure 6.19(b) two Zener diodes in series are included to reduce the voltage gain to unity when the signal becomes large.

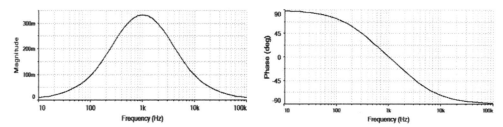

Figure 6.20. Frequency response of lead–lag filter.

Figure 6.21(a) is a simulated waveform from the Wien-bridge oscillator circuit shown in figure 6.19(a) with the following parameters: $R_1 = 10\,k\Omega$, $R_2 = 22\,k\Omega$, $R = 10\,k\Omega$, and $C = 10\,nF$. The voltage gain is 3.2 V/V. The simulation shows that the oscillation

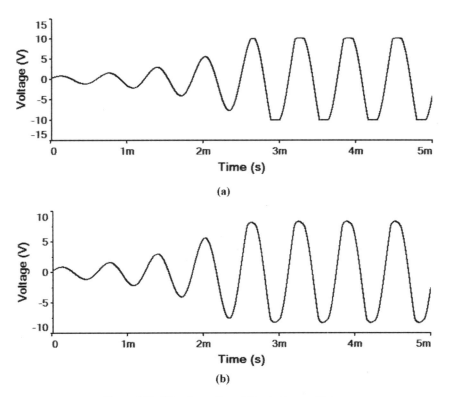

Figure 6.21. Waveforms from Wien-bridge oscillator.

starts quickly, but the output signal grows out of control and becomes saturated at the top and bottom. Keep in mind that there are upper and lower bounds for the output of op-amps, so the magnitude of the output signal cannot keep increasing.

Figure 6.21(b) is the simulated waveform from the circuit shown in figure 6.19(b) with an ideal op-amp. Now the amplitude of the output signal is under control and the saturation is avoided. Unfortunately, the distortion from a good sine wave is quite large, and thus Wien-bridge oscillators are no longer widely used any more.

William Hewlett—the co-founder of Hewlett-Packard Company—investigated Wien-bridge oscillators in his master's thesis in 1939. At that time semiconductor diodes were not available, so he used a filament to replace R_1 in the oscillator circuit. As we know, the resistance of metals gets higher with rising temperature. Therefore, R_1 becomes higher from ohmic heating when the output signal gets large, and this brings the gain down and then the output signal can eventually be stabilized. The first product of HP, the audio oscillator HP200A, was based on this design.

86214640R00066

Made in the USA
San Bernardino, CA
26 August 2018